George E Waring

The Sanitary Drainage of Houses and Towns

George E Waring

The Sanitary Drainage of Houses and Towns

ISBN/EAN: 9783744678759

Printed in Europe, USA, Canada, Australia, Japan

Cover: Foto ©berggeist007 / pixelio.de

More available books at **www.hansebooks.com**

THE SANITARY DRAINAGE

OF

HOUSES AND TOWNS.

SECOND EDITION, REVISED AND ENLARGED.

BY

GEORGE E. WARING, JR.
CONSULTING ENGINEER FOR AGRICULTURAL AND SANITARY WORKS.

"A hale cobbler is a better man than a sick king."

BOSTON:
HOUGHTON, MIFFLIN AND COMPANY,
The Riverside Press, Cambridge.
1881.

PREFACE.

THIS book has grown out of a series of articles originally published in the "Atlantic Monthly" magazine. The interest evinced in the subject by persons in every condition of life, and in all parts of the country, seemed to indicate that its more detailed treatment would be acceptable. As the investigations which the preparation of those articles made necessary have, together with professional studies, brought within my reach a wide range of material, I have presumed to submit what I have to say on the subject in this more permanent and more complete form.

The following chapters are not offered as of material value to such engineers and architects as have given attention to their subject, as these would naturally resort to the original authorities from which they have been so largely drawn. They are addressed more especially to the average citizen and householder, and are intended rather as an incentive to the securing of good work, than as a guide to the manner of its performance. For this reason they are largely devoted to the question of preventable disease; to the manner in which this increases our death-rate and lessens our worth to ourselves and to the community; and to those means which experience has shown to be best adapted for the removal of unhealthful conditions.

The instances are of course not rare in which the individual householder may have it in his power to secure, either by his own direction, or by his influence over the sanitary authorities of his town or city, an improvement of the conditions by which the health of his own household is now endangered. To this end I have endeavored to include so much of specific instruction as shall enable him to give a direction to the necessary local improvements.

Some of the points discussed are especially intended for the information of town authorities, and the usual sewer and health committees of local government boards.

If I shall have succeeded in inducing such persons to secure the needed reforms, and to insure their proper execution, this part of my work will have attained its best purpose.

NEWPORT, R. I., *March*, 1876.

This second edition is corrected and amended in accordance with the material recent improvements of the art.

NEWPORT, R. I., *June*, 1879.

CONTENTS.

CHAPTER I.
THE SANITARY RELATIONS OF DRAINAGE . . . . PAGE 9

CHAPTER II.
THE DRAINAGE OF HOUSES 71

CHAPTER III.
THE DRAINAGE OF TOWNS 105

CHAPTER IV.
ARRANGING PLANS FOR TOWN SEWERAGE 123

CHAPTER V.
THE CONSTRUCTION OF SEWERS 180

CHAPTER VI.
THE DETAILS OF HOUSE DRAINING 186

CHAPTER VII.
THE DRY CONSERVANCY SYSTEM 213

CONTENTS.

CHAPTER VIII.
VAULTS AND PRIVIES . . . . . .

CHAPTER IX.
LIERNUR'S PNEUMATIC SYSTEM OF SEWERAGE .

CHAPTER X
THE DISPOSAL OF SEWAGE . . . . .

CHAPTER XI.
THE DRAINAGE OF A VILLAGE . . . .

CHAPTER XII.
RECENT MODIFICATIONS IN SANITARY DRAINAGE

THE SANITARY DRAINAGE OF HOUSES AND TOWNS.

CHAPTER I.

THE SANITARY RELATIONS OF DRAINAGE.

"Mr. Wadley — described as a stout, robust gentleman — could not understand all the fuss made nowadays about the water question. Mr. Cooper cut the knot. He said that sin had brought disease into the world, and the Almighty permitted the outbreak of diarrhœa in their midst; neither doctors nor any one else could prevent it. Mr. Cooper is not far wrong. Sin has much to do with diarrhœa, especially municipal sin, which permits a population to drink sewage, and then coolly satisfies itself with referring the judgment to the Almighty."

THE art of Sanitary Drainage may almost be said to have been born — or reborn — but a quarter of a century ago, and it has contended with much difficulty in bringing itself to the notice of the public. Indeed, it is only within the past ten years that it has made its way in any important degree outside of purely professional literature.

Dr. Henry Maccormac says, "We live or we die, live well or miserably, live our full term or perish prematurely, accordingly as we shall wisely or otherwise determine."

Happily, men, — and women too, — are fast coming to realize the fact that humanity is responsible for much of its own sickness and premature death, and it is no longer necessary to offer an apology for presenting to public consideration a subject in which, more than in any other, — that is, the subject of its own healthfulness and the cleanliness of its own living, — the general public is vitally interested.

The evils arising from sanitary neglect are as old as civilization, perhaps as old as human life, and they exist about every isolated cabin of the newly settled country. As population multiplies, as cabins accumulate into hamlets, as hamlets grow into villages, villages into towns, and towns into cities, the effects of the evil become more intense, and in their appeal to our attention they are reinforced by the fact that while in isolated life fatal or debilitating illness may equally arise, in compact communities each case arising is a menace to others, so that a single centre of contagion may spread devastation on every side.

It is not enough that we build our houses on healthful sites, and where we have pure air and pure water; we must also make provision for preventing these sites from becoming foul, as every unprotected house-site inevitably must, — by sheer force of the accumulated waste of its occupants.

Houses, even of the best class, which are free from sanitary objections are extremely rare. The best modern appliances of plumbing are made with almost no regard for the tendency of sewer-gas to find its way into living-rooms, nor for other insidious but well known defects. So generally is this true, that it is hardly an exaggeration to say that unwholesomeness in our houses is practically universal. Hardly less universal is a curious sensitiveness on the part of the occupants of these houses to any suggestion of their short-comings.

Singularly enough, no one whose premises are subject to malarial influences seems willing to be told the truth with regard to them. No man likes to confess that his own well and his own cess-pool occupy the same permeable stratum in his garden; that the decaying vegetables in his cellar are the source of the ailments in his household; or that an obvious odor from his adjacent pig-sty, or from his costly marble-topped wash-stand, has to do with the disease his physician is contending against.

That the imperfections of our own premises are a menace to our neighbors is a still more irritating suggestion, and such criticism seems to invade the domain of our private rights. Yet surely there can be no equitable or legal private right whose maintenance endanger the well-being of others, — as our wide-spread disregard of the defects in our own houses does endanger the well-being of our fellow-townsmen. It is not possible, in a closely built town or compact neighborhood, for one to retain in his own grounds (either on the surface or in a vault or cess-pool) any form of ordure or festering organic matter, without endangering the lives of his neighbors, through either the pollution of the common air or the poisoning of wells fed from strata underlying the whole ground and more or less tainted by household wastes. Even if he might be permitted to maintain a source of injury to his own family, his neighbors may well insist that he shall not endanger them.

It being important for all that each be made to live cleanly, and the requirements of all, so far as the removal of the wastes of life is concerned, being essentially of the same character, the question of drainage is one in which the whole public is interested, and which should be decided and carried out by public authority, — so that all may have the advantage of the economy of organized work and the security of work well done. This applies not only to the construction of common sewers and public drains, but equally to those lateral branches of the public works which extend into private grounds and houses, — and *it applies to every detail of these.* So long as these important duties are left to the negligence of house-holders and to the demand for excreta among those who use manure, so long will they remain haphazard, unsatisfactory, and dangerous.

Wherever there is a systematically organized and well conducted board of health, it has been well suggested that their duties should include some power of veto upon the right of building houses upon unwholesome sites. All scavenging and disinfecting must, in order to be effective, be thorough and systematic, — which conditions can only be secured by the most careful public direction and supervision.

The drainage question is essentially a question of health and life. Dr. George Derby stated the whole case when he said: " The well are made sick and the sick are made worse for the simple lack of God's pure air and pure water." Air is infected and water is tainted, not only by defects in the public works, but quite as often and quite as dangerously by imperfections in household arrangements.

Neither Dr. Derby's statement nor the most perfect modern development of the art of cleansing towns by water-carriage has the merit of novelty. Hippocrates gave as the cardinal hygienic formula, " Pure air, pure water, and a pure soil," and after all these centuries we know nothing to add to it. Our modern sewerage works are thus far only taking us back to the cleanly condition of the most prosperous ancient cities; only lifting us out of the slough of plague-causing filth that marked the darkest periods of the Middle Ages; only continuing the wholesome revival that the Mohammedan Moors introduced among the unwashed Christians of Europe. It is a revival that has grown slowly, urged on by the harsh whip of disease and death. So late as the middle of the brilliant ninteenth century it had only begun to command the aid of the law, and as a subject of popular interest it can hardly yet be said to receive the attention of even the more intelligent members of society.

Yet when the subject is once considered, every thoughtful person must appreciate the fact that in seeking the advantages of community of living we necessarily depend at every turn upon our fellow-men, and that in this communion we lay ourselves open to the consequences of the neglect of others,— while we equally threaten others with the consequences of our own neglect. The influence of thoughtful persons cannot long be withheld from a movement whose object it is to popularize the knowledge of good and evil in the conduct of the daily life of the household and of society, and to make the public at large insist that each shall so regulate his action as to secure the greatest safety for all.

Public sanitary improvement is not the affair of the philanthropist alone, nor is the interest of the individual satisfied when he has made his own immediate surroundings perfect. Everything that can affect the health of the poorest and most distant of our neighbors may affect us; and, practically, the spread of disease in closely-built towns is more often than not, by the agency of public sewers, from the poorest classes upward, so that many a patient falling ill of contagious or infectious disease in the back slums of the city becomes the centre of a wide infection. The health of each is important to all, and all must join in securing it, — the public control, in the public interest, must extend to the sanitary condition of every household, — not among the poor alone, but at least equally among the rich. Indeed, from the greater complication of their plumbing work, the houses of the rich really require more careful supervision than do the simpler ones of the poorer class.

Not less important is the condition of the public sewers and drains.

An old-fashioned sewer has been well called a retort for the manufacture of sewer poisons which are "laid on" at every house by an ingenious system of pipes delivering an intermittent supply through every water-closet, bath-tub, and wash-basin, and producing its annual crop of zymotic disease.

The case has been very well stated by Dr. Sandwith. "Now, in doing away with the cess-pools all connected together, never properly cleansed, and last, and worst of all, communicating with the interior of dwelling-houses. If you have a case of typhoid fever a mile off, who knows but you may have the mysterious fungoid organisms conveyed into your own house through the water-closet. It is all very well to boast of traps and of similar mechanical arrangements, but remember there is such a thing as corrosion of metals, and the smallest defect, no larger than the interior of a straw, may introduce into your houses vast volumes of gas."

The great aim of all sewerage work is to secure to every member of the community his full supply of uncontaminated air, and, where wells are used, of pure drinking water.

Referring to the lower quarters of the city of Boston, Dr. Derby asks us to consider "what would be the effect upon the annual mortality in a community like Boston, if the wretched cellars and crystal palaces and rookeries and dens in which the extremely poor and improvident live could be depopulated, and their occupants transferred to well drained and lighted and ventilated buildings, of however cheap and simple construction; if all the foul fluids could be made quickly to depart by force of gravity through ventilated sewers; if all the foul solids could be removed without delay in carts provided with means for arresting putrefaction; if the blind alleys and narrow streets were opened to the admission of the air and of sunlight; if the old vaults were removed, the old cisterns torn down or filled, and the general principle of *cleanliness in its broadest sense* applied to air, water, and food." The picture would have been complete, had he suggested the well-known fact that the danger to the community from the class of diseases known as " pythogenic " (born of putridity) is not confined to those who live amid these filthy surroundings, but that the very sewers with which the better houses are drained are too often subterranean channels for conveying poisonous gases from the places of their origin to quarters which, without this transmission, might have their own drains so arranged that they would remain free from contamination.

Self-preservation is the first law of our nature; but it is a law which we ignorantly and constantly disregard in laying our life and health at the mercy of the foul conditions of life prevailing among our

SANITARY RELATIONS OF DRAINAGE. 17

neighbors, — and which we too often disregard in blindly trusting to the skillful but ill-advised work of our well-paid but untaught plumbers.

We roll up our eyes and stand aghast when contemplating the horrors of war; yet the mortality of war is trifling as compared with the mortality by preventable disease. England, in twenty-two years of continuous war, lost 79,700 lives; in one year of cholera she lost 144,860 lives.

We look idly on and see our population decimated by an infant mortality so great that its like among calves and colts would appall the farmer, and set the whole community energetically at work to discover a remedy.

It is estimated that for every person dying, twenty fall sick (Playfair estimates it at twenty-eight), and — to turn the argument in a direction best understood by many of our more influential neighbors — that every case of sickness is, on an average, equivalent to a loss of fifty dollars.

Dr. Stephen Smith says: "Man is born to health and longevity; disease is abnormal, and death, except from old age, is accidental, and both are preventable by human agencies."

Disease is not a consequence of life; it is due to an unnatural condition of living, — to neglect, abuse, or want.

Were any excuse needed for the constant reiteration of such truths as are known concerning the origin and spread of infectious diseases, it is to be found in the hope that by creating a public realization of the danger of sanitary neglect we may obviate the necessity that now seems to exist for the appearance of occasional severe epidemics, acting as scavengers and inducing the performance of sanitary duties whose continued neglect would lead to even more serious results.

Dr. Farmer speaks of pestilence as the angel "with which it would seem it has pleased the Almighty Creator and Preserver of mankind" to awaken the human race to the duty of self-preservation; plagues "not committing havoc perpetually, but turning men to destruction and then suddenly ceasing, that they may consider. As the lost father speaks to the family, and the slight epidemic to the city, so the pestilence speaks to nations."

The death-rate in the healthiest broad districts in England may be fixed at about fifteen per thousand per annum; but taking the whole kingdom into consideration, the death-rate is thirty-five per thousand, over one fourth of the deaths being due to preventable diseases. It is estimated that eighteen deaths take place every hour which might have been prevented by proper precaution. In addition to this, account must be taken of the lowering of the tone of health of those who survive, and of the existence of a vast number of weakly persons who are a tax on the community, and who transmit an inheritance of physical weakness to their offspring. Infants are most susceptible to unhealthful influences, and one half of the population of Great Britain dies before attaining the age of five years.

By another statement: "Looking at England as a whole, we see that of each one hundred persons who die, not quite ten have reached the standard old age of seventy-five years; and that of each one hundred children born, hardly seventy-four complete five years of life."

An ordinary epidemic any modern community will bear almost with indifference. The few who know the close relation between the disease and its preventable cause will generally maintain their accustomed indifference until their own circle is attacked, and even then they are powerless to arouse the authorities to the necessary action. It is only when an outbreak of more than ordinary malignity occurs that even the sanitary boards of most of our towns bestir themselves in the matter; but if the prevalence and the malignity be sufficient, there follows a most active cleansing of streets, purification of drains, and investigation of the private habits of the lower classes of the people. Then only is such attention given to the most obvious duty not only of the sanitary authorities, but of every man in the community, as, had it been exercised in advance, would have prevented every unnecessary death and every case of preventable sickness that has gone to swell the aggregate needed to attract public attention.

Nothing so arouses the alarm of a people as an epidemic of cholera; yet it is a singular fact that, even during the most severe cholera epidemics, the deaths from this disease are less than from many others which attract no attention and excite no apprehension. During the very alarming epidemic of 1849-50, there were 31,506 deaths from cholera in the United States. During the same period, there were more than this number of deaths from other diseases of the intestinal canal, and more from fevers alone.

That a proper use of known sanitary appliances is competent to remove the causes of a large class of fatal diseases is hardly disputed, and it is clearly proven by experience here and abroad.

Mr. Baldwin Latham, in his excellent work on "Sanitary Engineering," gives the following table, showing the effect on health of sanitary works in different towns in England:—

Name of Place.	Population in 1861.	Average Mortality per 1,000 before Construction of Works.	Average Mortality per 1,000 since Completion of Works.	Saving of Life per Cent.[1]	Reduction of Typhoid Fever Rate per Cent.	Reduction in Rate of Phthisis per Cent.
Banbury .	10,238	23.4	20 5	12½	48	41
Cardiff . .	32,954	33.2	22.6	32	40	17
Croydon .	30,229	23.7	18.6	22	63	17
Dover . .	23,108	22.6	20.9	7	36	20
Ely . . .	7,847	23.9	20.5	14	56	47
Leicester .	68,056	26.4	25.2	4½	48	32
Macclesfield	27,475	29.8	23.7	20	48	31
Merthyr .	52,778	33.2	26.2	18	60	11
Newport .	24,756	31.8	21.6	32	36	32
Rugby . .	7,818	19.1	18.6	2½	10	43
Salisbury .	9,030	· 27.5	21.9	20	75	49
Warwick .	10,570	22.7	21 0	7½	52	19

[1] It is to be remembered that even this great saving of life has been effected by works that are very far from perfect.

The average reduction of typhoid rate was nearly one half (47⅔ per cent.) in these twelve small towns. It is believed to be practicable, by the use of the most perfect known methods of drainage and ventilation, absolutely to prevent the occurrence of a single original case, and to confine all importations of the disease to the persons of those who bring them.

Two hundred years ago the death-rate of London was eighty per thousand; under the influence of sanitary improvements it has now been reduced to twenty-one and one-half per thousand, in spite of the enormous growth of the town and the great crowding to which many of its people are still subjected.

When the improvement of sewerage was actively undertaken in London some twenty-five years ago, it was found that the death-rate was so much reduced, in some of the worst quarters of the town, that if the same reduction could be made universal the annual deaths would be twenty-five thousand less in London, and one hundred and seventy-seven thousand less in England and Wales; or, by another view, that the average age at death would be forty-eight instead of twenty-nine, as it then was.

The early registration returns of England developed the fact that the prevalence of fatal diseases was in the case of some three times, of some ten or twenty times, and of others even forty or fifty times greater in certain districts than in others, and that these diseases raised the mortality of some districts from fifty to a hundred per cent. higher than that of other districts, the death-rate of the whole country being from thirty to forty per cent. above that of its healthiest parts.

The effect of sanitary improvement has been nowhere better shown than in the British navy, where in 1779 the death-rate was one in forty-two (this of able-bodied, picked men), and the sick were two in every five. In 1813, after the means and appliances of health had been furnished, the death-rate was one in one hundred and forty-three, and the sick two in twenty-one.

Less than a generation ago the idea prevailed that it was of doubtful propriety to ask why we are sick, and even to this day many believe that such an inquiry savors of irreligion. Happily this condition of otherwise intelligent minds is passing away.

While we know, thus far, comparatively little of the exact causes of disease, our knowledge at least points to certain perfectly well-established truths. One of these is that man cannot live in an atmosphere that is tainted by exhalations from putrefying organic matter, without danger of being made sick — sick unto death. It is true that not all of those who live in such an atmosphere either fall sick or die from its effects; but it is also true that not all who go into battle are shot down. In both cases they expose themselves to dangers from which their escape is a matter of good fortune. Fewer would be shot if none went into battle, and fewer would die of disease if none were exposed to poisoned air.

SANITARY RELATIONS OF DRAINAGE. 23

Our adaptability is great, and we accustom ourselves to withstand the attacks of an infected atmosphere wonderfully well; but for all that, we are constantly in the presence of the danger, and though insensibly resisting, are too often insensibly yielding to it. Some, with less power to resist, or exposed to a stronger poison, or finally weakened by long exposure, fall sick with typhoid fever or some similar disease, that springs directly from putrid infection. Of these, a portion die; the community loses their services, and it sympathizes with their friends in mourning that, "in the wisdom of a kind but inscrutable Providence, it has been found necessary to remove them from our midst."

In this way we blandly impose upon Divine Providence the responsibility of our own short-comings. The victims of typhoid fever die, not by the act of God, but by the act of man; they are poisoned to death by infections that are due to man's ignorance or neglect.

Pettenkofer states that, so far as the city of Munich is concerned, typhoid epidemics bear in their frequency or rarity a certain fixed relation to changes in the soil, which can only be surmised, but which correspond with the differences of elevation of the water-table, or line of saturation in the soil. The greatest mortality coincides with the lowest state of the water-table, and the least mortality with the approach of this to the surface of the ground.

Fifteen years' observation showed that the prevalence of typhoid was indicated by the water-level in the wells. This careful investigator believes that the cause of the disease exists not in the water, but in the soil; that it is due to certain " organic processes " in the earth.

The English investigators say that when the water in the well is low, its area of drainage is extended, and it draws typhoid poison from a greater distance.

Neither of these theories is inconsistent with the hypothesis that the disease is due to organic matter reaching the soil from house-drains, cess-pools, etc., and finally either carried into the well to poison the drinking-water to a degree that becomes apparent when, during drought, it is reduced to a small quantity, and its impurities are concentrated, or else left in the soil after the withdrawal of the water, and there exposed in such quantities to the action of the permeating air that poisonous gases are generated by their decomposition.

Professor S. W. Johnson, of the Sheffield Scientific School, at Yale College, in a paper on the Earth Closet, says: " The use of open vaults or water-closets emptying in cess-pools tends to fill up the soil with fæcal matter. A single vault poisons a circumscribed space around it. External to this limit the filth is destroyed by the action of the oxygen of the air, which is the great purifier. Within the limit named, the animal matters preponderate either constantly or at some period of the year,

They may long remain simply disagreeable without being dangerous, and may again, of a sudden, in a way whose details have as yet escaped investigation, become the seed-bed or the nursery of the infection that breaks out in fevers and dysentery. The danger increases as the quantity of filth and the number of its receptacles increase. To cover them up does not necessarily remove the evil. The putrid matters soak into the soil, and move upward and downward in it with the motion of the soil-water. When we have copious rains, they are carried down perhaps to nearly the level of the water in our wells. In the heat and drought of August, these matters rise again. In the absence of rain, the rapid drying of the surface creates an upward capillary flow of the ground-water. The matters which in rainy times follow the surface-water to the depths, in drought follow the ground-water to the surface."

It is very clear that no system yet applied has been so generally efficient in lessening and weakening the attacks of typhoid as the English system of water supply and impervious drainage, which gives drinking-water free from contamination, and leaves the air untainted by the decomposition of organic matters in the immediate vicinity of dwellings.

Whether the London theory or the Munich theory be correct, the general result of all investigations shows that typhoid fever stands in a certain relation to the amount of neglected filth permitted to poison water and air.

The Massachusetts Board of Health published in

1871 a copious report on the causes of typhoid as occurring in that State. It concludes that the propagation of the fever is occasioned by a poison " as definite as that which causes vaccine disease ;" and divides the means of propagation under three heads: first, drinking-water made foul by the decomposition of any organic matter, whether animal or vegetable, and specially by the presence in such water of excrementitious matters discharged from the bodies of those suffering from typhoid fever; second, propagation by air contaminated by any form of filth, and specially by privies, cess-pools, pig-sties, manure-heaps, rotten vegetables in cellars, and leaky or obstructed drains ; third, emanations from the earth, occurring specially in the autumnal months and in seasons of drought.

The agency of tainted water was enunciated by Canstatt, in Germany, in 1847, and many later medical writers have confirmed the theory.

So far as Massachusetts towns are concerned, the contamination of wells, though a prominent, was not found to be a preëminent cause of typhoid ; numerous instances show this to have been active, but other causes, such as foul drains, sewer-gas, etc., are more important. It appears that the attack is more frequently received through the lungs than through the intestines. While it may be necessary that a marked quantity of impurity should exist in drinking-water before it can do us harm, an extremely small proportion of impurity in air is greatly to be apprehended, for we drink but a comparatively small amount of water, while we inhale, every twenty-four hours, from one to two thousand gallons of air. At the same time, the evidence of the communication of disease by tainted drinking water is strong and unmistakable, not only in Massachusetts, but elsewhere.

There has recently been an excitement in London concerning the condition of town pumps, especially of the celebrated Aldgate pump in the city, which has finally been ordered closed.

The following quotation,[1] describing the condition of this pump, is given in illustration of what has so often been said concerning the peculiarly pleasant character of well water to the influence of which disease has been distinctly traced : —

" The purity of the water of Aldgate pump is a firmly rooted tradition in the minds of many of the citizens, and especially of the poorer denizens of the East End. So great a hold, indeed, has this belief, that a great many of them consider that a morning draught of spring water from Aldgate pump is a sovereign remedy for many ailments, and send for the water very religiously when they feel ' out of sorts.' There is much in the flavor and appearance of the water which explains this belief. It is clear, sparkling, and has that cool saline flavor which is so very agreeable to the palate, and which is harmless enough when the saline ingredients are not accompanied by organic taint. In this case, however, the analysis which we have had made by Pro-

[1] *Sanitary Record.*

fessor Wanklyn, and of which we last week gave the details, shows that this cool refreshing flavor is due to the impregnation of the water with salts derived from decomposing sewage which evidently finds its way into the well, partially filtered and decomposed by the surrounding soil. The soil itself is evidently loaded with organic matter, and does not form an efficient filter; thus the water is polluted with a considerable amount of organic matter."

Anent this pump, Punch had the following squib at the time of the recent panic concerning Turkish and Egyptian bonds: —

"GENERAL SHUT UP.
"(Aldgate pump included.)
"O'er 'Change still hangs the fatal spell;
Clerics and spinsters Turkish sell;
Egyptian drafts, too, downward jump,
And none may draw on Aldgate pump."

Mr. William Eassie, writing upon the sanitation of houses, says: —

"The author had occasion lately to suspect, from its very sparkling character, the water taken from a well in a very healthy looking position, which supplied several families and a large dairy, and had the water analyzed. Professor Wanklyn's report upon it was that it was absolutely poisonous, and then it was found that the constant drinkers of this water had long been suffering from a skin disease. Inquiry also revealed that a farm steading had formerly stood there, and we can guess what these used to be in the olden time, and that the subsoil was, therefore, full of impurities. No amount of filtering could render water of this kind pure. Filters can be made useful, without doubt, but even they must be examined periodically, and a proper material chosen. He had an analysis made of some water drawn at the butler's pantry in a nobleman's house, and it was proved that the filter in use, with its impure animal charcoal contents, actually rendered a pure supply from the mains unfit to drink."

There is reason to suspect the poison to be sometimes, if not quite generally, odorless, and the danger seems to be the greatest where the natural process of decomposition is secluded from air and light. The decay of vegetables in dark and unventilated cellars, and of house wastes or street wash in unventilated sewers, are especially to be feared.

In the town of Pittsfield, when the board of health assiduously attended to the removal of all nuisances, there was a very decided falling off in the number of cases of typhoid fever.

Derby says: "Whether the vehicle be drinking-water made foul by human excrement, sink drains, or soiled clothing, or air made foul in inclosed places by drains, decaying vegetables or fish, or old timber; or, in open places, by pig-sties, drained ponds or reservoirs, stagnant water, or accumulations of filth of every sort,— the one thing present in all these circumstances is decomposition."

If anything has been clearly proven with reference to the whole subject, it is that nearly all of the causes of typhoid fever are strictly within human control.

Dr. Grimshaw regards simple fever as an abortive attempt at either typhus or typhoid, the one arising from decomposing sewage and the other from over crowding, and thus ascribes even slight fevers almost invariably to unsanitary conditions. A Warwickshire medical officer of health firmly believes that typhoid is far from wrongly named by Dr. Murchison, "pythogenic," for that it cannot only be fostered, but be produced, *de novo*, by decomposing organic matter, and in most if not all of Dr. Armistead's cases, foul water or other unsanitary surroundings appear to have been present in greater or less amounts.

Dr. Benjamin Rush, an eminent physician of the last century, was so satisfied that the means of preventing pestilential fevers are " as much under the power of human reason and industry as the means of preventing the evils of lightning or common fire," that he looked for the time when the law should punish cities and villages "for permitting any sources of malignant or bilious fevers to exist within their jurisdiction."

No dense population can hope to escape recurrent pestilential diseases, unless the inhabited area is kept habitually free from the dejections and other organic wastes of the population.

The instance of the "Maplewood" young ladies' school, at Pittsfield, Massachusetts, has been so often quoted in sanitary writings during the past ten years, that it must seem almost an old story to those who are familiar with the literature of the subject; but it is at the same time such a striking instance of the possibilities of the evils with which we are contending, that it can never lose its interest, and it is to be hoped that it may always remain the worst instance of its sort in our country's record.

The house was a large one, built of wood, closely surrounded by trees, with a foul barn-yard near it, containing water in which swine wallowed, and emitting offensive odors. The cellar of the centre main building was used for storing vegetables, and its private closets connected by closed corridors with the main halls of the building. The kitchen drain opened eighty or ninety feet away from the building. The vaults of the private closets were shallow and filled nearly to the surface with semi-fluid material (they received the slops from the sleeping-rooms). The house seems to have been beset with danger on every side, and it was often necessary, in the heat of summer to close the windows, to keep out offensive odors. The whole case was examined after the attack by Drs. Palmer, Ford, and Earle, of the Berkshire Medical College, and they took, so far as possible, the testimony of every member of the household and of the relatives of those who had died after being removed to their homes. Their investigation fixed the origin of the Maplewood fever (which was clearly marked typhoid) unquestionably upon the unhealthfulness of the air of the house, made impure by the causes above specified.

This Maplewood fever is one of the most fatal ever recorded. Of seventy-four resident pupils heard from, sixty-six, or nearly ninety per cent., had illness of some sort, and fifty-one, or nearly sixty-nine per cent., had well-marked typhoid fever. Of the whole family of one hundred and twelve persons, fifty-six had typhoid fever, and of these fifty-six, sixteen died. These proportions applied to the eight thousand people living in Pittsfield would have given four thousand cases of typhoid fever within the space of a few weeks, and of these eleven hundred and forty would have died. The outbreak was, however, so entirely local, that some physicians in Pittsfield had no cases, and others only two or three. The Maplewood fever was a sudden explosion. It broke up the school in a very short time, and the pupils scattered to their homes, where, under the influence of pure air, many recovered.

Dr. Palmer says of this epidemic, "Before the investigation, the matter was spoken of as the act of a mysterious Providence, to whose rulings all must submit. Looking with the eye of science upon the overflowing cess-pools and reeking sewers as inevitable causes, and with the eye of humanity upon the interesting and innocent victims languishing in pain and peril, or moldering in their shrouds, I could but regard such implications of Providence, though perhaps sincerely made, as next to blasphemy, especially when uttered by the agents who were to be held responsible; — though the prayer of charity might have been, 'Father, forgive them, they know not what they do.'"

The sanitary reforms recommended by the examining physicians being carried out, Maplewood became, and still remains, free from malarial disease.

Dr. John L. Leconte, in reporting his inspection of a school in Burlington, New Jersey (St. Mary's Hall), where there had been a serious outbreak of typhoid, says that the water supply was taken from two rain-water cisterns, in building which, as their bottoms were below the level of the soil water, a hole was left open to relieve the pressure while the cement was hardening. These holes were plugged, and the water supply was made to depend entirely upon the river. A year later, the plugs were removed, bringing the cisterns into communication with the soil water. Some time afterwards, privy vaults were dug, one of them only ten or twelve feet from the cisterns, and although these were supposed to be tightly made, the soil (after three years) became poisoned with the effluvia and infiltrations, and the water from the cisterns became contaminated. The disuse of these cisterns was advised on the 18th of December, 1874, the water being taken directly from the river. Ten days later, the last case of typhoid fever occurred, and the school has since that time been quite free from all similar diseases.

Although many of the pupils were attacked, the servants escaped entirely, and it was found that they had drank no unboiled water, using only tea and coffee. Among the pupils it was found that of seven who drank only water, six had been attacked with typhoid.

Dr. Leconte makes the following capital recommendations for the prevention of zymotic diseases: —

"1. Before the plans of the building are fully matured, let an expert in sanitary studies be employed to give directions to the architect in all that relates to ventilation, drainage, and water supply.

"2. After the building is completed, no alterations should be made affecting these three essentials of good hygienic condition without the suggestion of a practiced sanitarian.

"3. There should be stated inspections, say twice a year, of each institution by some sanitarian of acknowledged merit, who, after close examination and the correction of any defect, would give a certificate to be published in the circular or announcement of the school.

"4. On the outbreak of any zymotic disease in the institution, the advice of a sanitarian expert should at once be obtained, in order that means may be taken for its restriction, suppression, and prevention."

A century ago epidemic diseases carried with them only calamity, not culpability; but now, when their occurrence is chargeable to willful ignorance or to wicked neglect, Dr. Rush's prophecy should be fulfilled and the law should hold the community responsible for every death permitted to occur from preventable disease within the area that it controls.

. Dr. Anstie, in his "Notes on Epidemics," after describing the fouling of wells by the escape of cesspool matter, and the fouling of the interior air of houses by reason of imperfect drain-traps, says: —

"In short, all observers arrived at the conclusion that it would be possible, by rendering our drinking water absolutely pure, and by disinfecting our sewage at the earliest moment, entirely, or almost entirely, to suppress typhoid fever."

Dr. John Simon, in his Report of 1874 (as medical officer of health to the Privy Council), says: —

"Whether the ferments of disease, if they could be isolated in sufficient quantity, would prove themselves in any degree odorous, is a point on which no guess needs be hazarded; but it is certain that in doses in which they can fatally infect the human body they are infinitely out of reach of even the most cultivated sense of smell, and that this sense (though its positive warnings are of indispensable sanitary service) is not able, except by indirect and quite insufficient perceptions, to warn us against risks of morbid infection."

. . . .

"The ferments, so far as we know them, show no power of active diffusion in dry air: diffusing in it only as they are passively wafted, and then, probably, if the air be freely open, not carrying their vitality far; but, as moisture is their nomal medium, currents of humid air (as from sewers and drains) can doubtless lift them in their full effectiveness, and, if into houses or confined exterior spaces, then with their chief chances of remaining effective ; and ill-ventilated, low-lying localities, if unclean as regards the removal of their refuse, may especially be expected to have these ferments present in their common atmosphere, as well as of course teeming in their soil and ground-water.

" Medical knowledge in support of this presumption has of late been rapidly growing more positive and precise ; and at the moment of my present writing, I have the gratification of believing that under my Lords of the Council it has received an increase which may be of critical importance, in a discovery which seems to give us for the first time an ocular test of the contagium of enteric fever : in the discovery, namely, of microscopical forms, apparently of the lowest vegetable life, multiplying to innumerable swarms in the intestinal tissues of the sick, penetrating on the one hand from the mucous surface into the general system of the patient, and contributing on the other hand, with whatever infective power they represent, to the bowel-contents which have presently to pass forth from him. Adverting then summarily, in an administrative point of view, to the present state of medical knowledge and opinion as to the way in which enteric fever spreads its infection in this country, I would say that it is difficult to conceive, in regard to any causation of disease in a civilized community, any physical picture more loathsome than that which is here suggested; that apparently, of all the diseases attributable to filth, this, as an administrative scandal, may be proclaimed as the very type and quintessence; that, though sometimes by covert processes which I will hereafter explain, yet far oftener in the most glaring way, it apparently has an invariable source in that which of filth is the filthiest; that apparently its infection runs its course, as with successive inoculations from man to man, by instrumentality of the molecules of excrement which man's filthiness lets mingle in his air, and food, and drink."

Dr. Austin Flint says: " Typhoid fever is very rarely if ever communicated by means of emanations from the bodies of patients affected with the disease. It does not spread from cases in hospitals to fellow-patients, nurses, and medical attendants. Isolated cases are numerous, occurring under circumstances which preclude the possibility of contagion. Its special cause may be a product of the decomposition of collections of human excrement."

Dr. Simon, speaking of the action of infective matters on the mucous membrane of the intestinal canal, says: —

" Whether they have been breathed, or drunk, or eaten, or sucked up into the blood-vessels from the surface of foul sores, or directly injected into blood-vessels by the physiological experimenter, there peculiarly the effect may be looked for; just as wine, however administered, would ' get into the head,' so the septic ferment, whencesoever it may have entered the blood, is apt to find its way thence to the bowels, and there, as universal result, to produce diarrhœa." He believes that typhoid fever is a "specific diarrhœa," and that every discharge from the bowels of the patient must teem with the contagium of the disease.

Dr. Flint investigated an outbreak of typhoid fever in a village in Western New York, in 1843. No case of typhoid fever had ever been known in the county. The community numbered forty-three persons; twenty-eight of these were attacked with fever, and ten died. All of those affected obtained their drinking water from a well adjoining the tavern; but one family, living in the midst of the infected neighborhood, owing to a feud with the tavern-keeper, did not use this water, and escaped infection. Two families lived too far away to use this well. This immunity on the part of the enemy of the tavern-keeper led to a charge that he had maliciously poisoned the well, a charge which resulted in a suit for slander and the payment of one hundred dollars damages. At this time the idea that typhoid fever might be communicated by infected drinking-water had not been advanced, but its truth receives strong confirmation from the fact that a passenger, coming from a town in Massachusetts where typhoid prevailed, and traveling westward in a stage-coach, having been taken ill, was obliged to stop at this tavern. Twenty-eight days after his arrival he died of typhoid fever, and thus, doubtless, communicated in some way to the water of this well the germs of the disease, which speedily attacked every family in the town, except the three which did not resort to it for their supply. Dr. Flint states it as his opinion that "in typhoid fever the contagion is in the dejections, and this fever may be, and generally is caused by a morbific matter produced in decomposing excrement from healthy bodies." And he believes that "the spontaneous occurrence of this disease is to be avoided by a complete precaution against the pollution of water or air by the dejecta from healthy persons."

In the summer vacation of 1874, ten students from Oxford went on a reading party to a rural retreat in Cornwall, which was recommended as of undoubted healthfulness and of quiet seclusion. They fell into a fever trap. The water and the soil of this village were polluted until it equaled the worst slums of Liverpool. Detecting the sanitary short-comings of their retiring-place, they beat a hasty retreat, but they carried with them the germs of the disease, and before many days six of the party were down with typhoid fever; one has since died.

Dr. Alfred Haviland gives an instance in which in Uppingham, in England, an epidemic of typhoid fever originated in a house of the best class:—

"Though the house itself in which the fever first showed itself is a splendid mansion, the architect seems to have altogether forgotten to provide for the health of its inmates. Gigantic cess-pools were in close relation to the water supply, and every arrangement was made for the pollution of the air by regurgitation of gases from the water-closet."

The local government board of England lately deputed Dr. Thorne to investigate an outbreak of typhoid at Brierly. He found that the spread of the fever was due to the poisonous dejecta of the patients. Wherever those dejecta went, poison and disease went also. The original case was in the person of a dairy-man, and was of a mild type; but it was followed by two other cases in the same house, and, by the tainting of the milk vessels, the infection was carried to thirty-eight houses in the village, in twenty-three of which the fever appeared. From these centres it spread by excremental contamination until nearly the whole village was attacked. Dr. Thorne " wished it to be distinctly understood that he by no means attributed all the cases occurring to the use of milk from the infected dairy; for when once the disease was started another powerful means for distributing it came into operation;" and he proceeds to show a very defective condition of the vaults and drains. His irresistible conclusion was that the outbreak had been due, primarily, to the use of milk from an infected dairy, and that bad drainage and bad disposal of excrement had done the rest.

Dr. Duncan, in his work on typhoid fever, speaks of Crosshill, a suburb of Glasgow, where for three years preceding 1875 the average death-rate was only seventeen per thousand; that of the city itself during the same period being thirty per thousand.

In 1874 there was no death from typhoid in Crosshill. From January 18 to April 20, 1875, there were twenty-four deaths in connection with this epidemic. From January 18, when the epidemic began, until March 31, when it ended, there had occurred two hundred and eighty cases in Crosshill and its neighborhood. This outbreak was distinctly traced to milk coming from a farm where the family was down with typhoid fever. Dr. Duncan attended sixty-eight cases, sixty-four of which could be traced directly to the tainted dairy; the other four were smitten late in the epidemic, and had been visiting and drinking in infected houses.

During the autumn of 1874 there was an outbreak of typhoid fever in the town of Lewes, about four hundred and fifty cases occurring. The town is divided into three sections, each having its own water supply, and the disease was confined almost entirely to the division supplied by the Lewes Water Works Company. This company furnished an intermittent supply of water, the head being turned on for three or four hours in the morning and for the same time in the afternoon. When the head is taken off, the pipes empty themselves, sucking in air at every opening. Examination showed that there were many water-closets, some of them used by fever patients, which were supplied by pipes leading directly from the water-mains into the soil-pan, and that it was a common habit to leave the taps open so that the closets should be flushed whenever the water was turned on. There were leaks in some of the old mains, and many of these were laid in soil fouled with the overflow of vaults. In one case a leak was found in a water-main where it passed through a sewer. The lead service-pipes of houses were frequently honey-combed, especially in the immediate vicinity of vaults, and in one case a leak was found directly under a vault. In seeking for this while the water was subsiding in the mains, the opening was exposed, and the whole contents of the vault were sucked into the water-pipe. In short, on every occasion of the subsiding of the water supply, air was drawn in violently at every opening, and the pipes thus received air contaminated by closets and vaults, and air from within a public sewer; indeed, in some cases at least, particles of excrement were drawn in from closet pans. In one section of the town only sixty houses out of a total of four hundred and fifty-four were supplied by the water-works company, and in this section, with the exception of two infants, every case of typhoid fever occurred in these sixty houses, to the total exclusion of the other three hundred and ninety-four. Even after the epidemic became rife, and there were many other means for its extension, it was found that twenty-seven per cent. of the town-water houses had been attacked, and only six per cent. of all the others.

There has recently been an investigation into the origin of an outbreak of "filth fever" in Over-Darwen, England, the origin of which for a long time eluded the careful search of the authorities. It was finally worked out by a sanitary officer dispatched from London. The first case was an imported one, occurring in a house at a considerable distance from the town. The patient had contracted the disease, came home, and died with it. On first inquiry it was stated that the town derived its water supply from a distance, and that the water was brought by covered channels and could not possibly have been polluted by the excreta from this case. Further examination showed that the drain of the closet into which the excreta of this patient were passed emptied itself through channels used for the irrigation of a neighboring field. The water-main of the town passed through this field, and although special precautions had been taken to prevent any infiltration of sewage into the main, it was found that the concrete had sprung a leak and allowed the contents of the drain to be sucked freely into the water-pipe. The poison was regularly thrown down the drain, and as regularly passed into the water-main of the town. This outbreak had a ferocity that attracted universal attention; within a very short period two thousand and thirty-five people were attacked, and one hundred and four died. The report of this investigation closes as follows: " If an inquest were held on every case of death from typhoid fever, as we have long contended there should be, a similar relation of fatal effect to preventable cause could nearly always be traced, and may always safely be presumed."

Thus much attention has been given to the subject of typhoid fever because it is universally recognized as the typical pythogenic disease, and the most prominent of those which are believed to be entirely preventable by human agency.

The recent great prevalence of a very fatal form of diphtheria in New York, under conditions which seem to connect its origin with the escape of sewer gas into houses, brings it conspicuously into the same class.

Two other prevalent scourges, not ascribed to organic uncleanliness but connected with the question of soil-water removal,— consumption and fever and ague, — must have a prominent place in any discussion of sanitary drainage.

The scientific world has been quick to accept the suggestion of Dr. Bowditch, of Boston, that the genesis of pulmonary disease seems often to be connected with excessive moisture either arising from a wet soil, from a clay subsoil, which is usually a cause of damp and cold, from springs breaking out near the site of the house, from sluggish drains, damp meadows, ponds of water, and other sources of fog and atmospheric moisture, or from too close sheltering by trees. To one or more of these causes it is now thought that we may ascribe the origin of a large proportion of the cases of that painful disease which, more than any other, characterizes New England.

Dr. Bowditch says, " Private investigations in

Europe and America have in these later times proved that residence on a damp soil brings consumption; and second that drainage of the wet soil of towns tends to lessen the ravages of that disease."

In 1865-66 the British government instituted an examination into the effect of drainage works on public health. Twenty-four towns sewered by the modern system were examined. "It appeared that while the general death-rate had greatly diminished, it was most strikingly evident in the smaller number of deaths from consumption." As Bowditch has pointed out, the drying of the soil as an incidental effect of sewerage had led to the diminution of this disease.

Those ailments which are caused by the influence of stagnant water, or excessive wetness of the soil — consumption in its most fatal form being one of them — may be much alleviated by the simple removal of the drainage-water, through exactly the same process that is employed in farm drainage.

The connection of fever and ague with soil moisture, and with the obstructed decomposition of vegetable matter in saturated ground or in moist air, is almost universally recognized.

The improvement resulting from drainage is fully attested by wide areas in England, where whole neighborhoods have been drained for farming purposes, and where, as a consequence, malarial diseases have entirely disappeared.

In the report of the Staten Island Improvement

Commission (1871), it is stated that where the foundations of the dwelling and the land about it for a certain space have been thoroughly underdrained, and where considerable foliage interposes between such space and any exterior source of malaria, the liability to disease is greatly reduced, and there is little danger that fever and ague would be contracted by the inmates of such a house, except by exposure outside their own grounds. An instance is cited where four adjoining farms, near Fresh Kills, were drained. Close to each of these farms there has been much malarial disease, but the seventy people living on them have had scarcely a symptom of it. In another quarter formerly very malarial, the occupants of which carried to other residences the disease there contracted, those who remained after the thorough drainage of the land have recovered, and have not suffered at all since; while those who moved to them after their drainage have lived there for years without injury. In this case as in the first, the neighborhood beyond the influence of the under-drains is still highly malarial.

Pulmonary diseases, especially the early stages of consumption; all continued fevers, especially typhoid fever; degenerative diseases, such as scrofula and cancer; and uterine diseases, both of tissues and of function, are stated by the Staten Island Commission, to become less severe with the natural or artificial reduction of the level of the ground-moisture.

The Secretary of the General Board of Health

(England) published in 1852 "Minutes of Information, collected in respect to the drainage of the land forming the sites of towns, etc."

He says: "When experienced medical officers see rows of houses springing up on a foundation of deep, retentive clay, inefficiently drained, they foretell the certain appearance among the inhabitants of catarrh, rheumatism, scrofula, and other diseases, the consequence of an excess of damp, which break out more extensively and in severer forms in the cottages of the poor, who have scanty means of purchasing the larger quantities of fuel, and of obtaining the other appliances by which the rich partly counteract the effects of dampness. Excess of moisture is often rendered visible in the shape of mist or fog, particularly towards evening. An intelligent medical officer took a member of the sanitary commission to an elevated spot from which his district could be seen. It being in the evening, level white mists could be distinguished over a large portion of the district. "These mists," said the officer, "exactly mark out and cover the seats of diseases, for which my attendance is required. Beyond these mists, I have rarely any cases to attend, but midwifery cases and accidents." Efficient drainage causes the removal, or at least a diminution of such mists, and a proportionate abatement of the disease generated or aggravated by dampness.

"After houses built in the manner described, have been inhabited for some time, and especially if crowded, fevers of a typhoid type are added to the preceding list of diseases, in consequence of emanations from privies and cess-pools. The poisonous gases, the product of decomposing animal and vegetable matter, are mixed with the watery vapor arising from excessive damp (such vapors being now recognized as the vehicle for the diffusion of the more subtle noxious gases), and both are inhaled night and day, by the residents of these unwholesome houses. A further consequence of the constant inhalation of these noxious gases, which have an extremely depressing effect, is inducing the habitual use of fermented liquors, ardent spirits, or other stimulants, by which a temporary relief from the feeling of oppression is obtained."

In the English Sanitary Report for 1852, the following propositions are laid down : —

"1. Excess of moisture, even on lands not evidently wet, is a cause of fogs and damps.

"2. Dampness serves as the medium of conveyance, for any decomposing matter that may be evolved, and adds to the injurious effect of such matter in the air; in other words, the excess of moisture may be said to increase or aggravate excess of impurities in the atmosphere.

"3. The evaporation of the surplus moisture, lowers temperature, produces chills, and creates or aggravates the sudden and injurious changes of temperature, by which health is injured."

The copious evidence taken by the Metropolitan Sanitary Commission, in 1848, concerning the effect of ordinary agricultural land-drainage, as practiced in England, upon the improving healthfulness of men and animals, and upon climate, resulted in the production of a vast amount of evidence of the most telling character, to review which, even briefly, would be impossible in this limited space; but it clearly showed that all the benefits that the advocates of land-drainage have claimed for it, had already been fully sustained by English experience.

The agricultural drainage of the land in and about the sites of towns, and the soil-drainage which is usually effected, even where no especial provision is made for it, by the ordinary works of sewerage, has fully demonstrated the sanitary benefit arising from the removal of stagnant water, or water of saturation, from the soil. The earth acts upon foul organic matters much in the same way that charcoal would do, having, though in less degree, the same sort of capacity for condensing within its pores the oxygen needed to consume the products of organic decomposition. But no soil can act in this way so long as its spaces are filled with water, and in order to make it an efficient disinfectant it is necessary to withdraw its surplus moisture and thus admit atmospheric air within its pores.

It is now generally believed that in addition to the many other evils of excessive soil-moisture, its effect in rendering a dwelling-house cold and unwholesome is especially marked in encouraging the formation of tubercles in consumptive subjects; and the various forms of malarial fever, neuralgia, influenza, dysentery, and diseases of the bowels, are thought to be aggravated by excess of moisture in the soil immediately about human habitations.

During the past thirty or forty years very large contiguous areas have been drained in England for agricultural purposes, and an invariable result of the improvement has been a great decrease of malarial diseases, such as fever and ague and neuralgia. The vast fen-lands of Norfolk, Lincolnshire, and Cumberlandshire were formerly the seat of very wide-spread diseases of a malarial type. Since the drainage of the fens these diseases have become comparatively rare and mild in form; and it is asserted with regard to England generally, that such diseases " have been steadily decreasing both in frequency and severity for several years; and this decrease is attributed in nearly every case mainly to one cause — improved land-drainage."

The well-known Mr. James Howard, of Bedford, England, says, " In my own county, ague and fever thirty or forty years ago were very common in certain villages; since draining has been carried out the former has quite disappeared, and the latter has greatly decreased."

So far as the question of social prosperity is concerned, it is quite proper to consider the financial aspects of the question of health. The body politic has perhaps no compassion for the sufferings of the poor invalid or the bereaved mourner, but it has a quick and a vital interest in everything affecting its worldly prosperity, and the deepest foundation of its worldly prosperity lies in the strength and efficiency of its members.

Dr. Boardman of Boston, in the sixth annual report of the Massachusetts Board of Health, enters into a calculation, based on numerous data, which seem to be sufficiently proved.

In the metropolitan district, including Boston, the average loss of time from sickness for each individual is twenty-four days in the year. In the western district, including Berkshire County, the loss is about fourteen days; and the average for the whole State is nearly seventeen days for each member of the population. This was in 1872; a similar computation for the previous eight years shows an average of fourteen days for each person. Calculating the cost of nursing, medical attendance, etc., and the loss of time to persons of a productive age, he finds that the loss to the State from the sickness of working people alone is over fifteen million dollars; and the same computation for the entire population would amount to nearly forty million dollars.

Assuming that out of the nineteen persons in every thousand who die annually in the whole State of Massachusetts, four might be saved by the avoidance of preventable diseases, — and this is certainly very low, for it may be reasonably assumed that eleven per thousand is the *natural* death-rate, or the lowest that can be attained, — the following saving to the State would result: the annual mortality being 26,619 with a death-rate of nineteen per one thousand, it would be, with a death-rate of fifteen per thousand, 21,015, or an annual saving of 5,604 lives. Good grounds are given for assuming that every death represents a total of 730 days of sickness and disability; the aggregate saving from sickness therefore would be 4,090,920 days, which would amount to $8,181,840, or for the working population alone $3,190,916, which latter sum would represent the interest on more than fifty million dollars. The practical question then which the commonwealth should consider is whether an investment of fifty million dollars in the improvement of the sanitary condition of its population, and in their enlightenment as to means for preserving health, would result in a reduction of the death-rate from nineteen to fifteen. If it would do so, then the investment would be a profitable one. That it might do this, and even more, is proven by English experience, and no one can doubt it who will give even casual attention to the degree to which human life, in even our best communities, whether in town or country, is hourly endangered by the unwholesome conditions under which it exists.

In every household in which a pronounced case of typhoid occurs, it may fairly be assumed that the value of the whole family to themselves and to the community is distinctly lessened; and the large proportion of "debilitated and weakly" persons

SANITARY RELATIONS OF DRAINAGE. 53

found in all our communities are but half-way victims struggling against the assaults of foul air and contaminated water. Their lives are permanently dulled by a malaria they are in part able to withstand.

In this ever-waged battle there are wounded as well as killed; and in England it is recognized that "convulsions" and many attacks of nervous ailments are marks of excremental poisoning.

There are several diseases which are now known to indicate more or less definitely unfavorable sanitary arrangements, and as the knowledge of hygiene extends, other diseases are being added to the list. Nervous toothache, neuralgia, scarlet fever, cholera, dysentery, diphtheria, cerebro-spinal meningitis, and consumption are among those which are either generated by foul air or foul water, or which are made worse because of unhealthy surroundings.

The "New York Medical Record" of June 19th, says, "It may safely be said that diphtheria, does not appear to have any connection with the distribution of the old water-courses of this city; also that a large number of cases have originated, without any suspicion of contact with the diseased matter in any form, while in some of these instances, sanitary defects of a very serious kind, have been found in the dwellings, making it highly probable that noxious emanations and the like have produced the disease. It may possibly be due to the foul emanations from slaughter-houses, and other nuisances, or it may arise from some accident or neglect in one's own dwelling, where it was supposed that every sanitary regulation was vigorously enforced. Such might be the explanation of the following interesting case : A prominent physician of this city was suddenly taken ill of diphtheria, and was confined to his room for five days. On recovering, and making a careful inspection of his premises, he found that in some unknown way the soil-pipe carrying the waste from the adjoining houses, had burst, letting in upon his cellar-floor, a collection of rottenness and filth, that was of the most disgusting kind. It is difficult always to be able to make a careful examination of the premises in this way, but it seems probable that if each case of diphtheria were carefully investigated, a large number of the so-called idiopathic cases, might be traced to some such source. Many similar instances probably occur to the mind of most practitioners, and there seems to be no reason why such influences as these described, should not in many cases be causes of diphtheria, just as they may often produce typhoid fever, puerperal fever, and erysipelas, — an opinion that is beginning to be very generally held."

The following is taken from the "Sanitary Record" of March 13, 1875 : —

"In consequence of an outbreak of diphtheria in Homsey, Mr. Oakeshott, the medical officer of health for the district, instituted inquiries, and traced the cause to the escape of sewer-gas into houses. The first case occurred to a child attending a small school. The house was disinfected, and it was supposed that the disease had been stamped out, but several other cases having occurred, the sanitary inspector made a minute examination, and found that notwithstanding the house was generally in a good condition, the drains had recently been connected with the main sewer, and since then foul smells had been complained of. The traps to the sink in the kitchen, where the school was held, were defective, and, on measuring the velocity of the rush of sewer-gas from the sink, he found it to be two to three cubic feet per minute. The room was only about ten feet square. It was consequently very bad for the children there, and the intensity of the poison rapidly proved fatal, as might be expected, but it seemed impossible that thirty cases could have occurred, through these few children who first had the disease. On examining the Fortis Green National Schools, Mr. Oakeshott and the sanitary inspector found a pit at the rear, full of foul soil, the stench being very bad. This was quite enough to cause the later outbreak. Mr. Forstall, medical officer of health of Highgate, who had been referred to, stated that in three cases of diphtheria which he attended in one family, sewage was found to have percolated under the floor. He attributed the outbreak which occurred at Fortis Green, to sewage gas. Great complaints had been made of the foul smells emanating from the main sewers; the prevailing opinion being, that the smells were worse since the completion of the drainage scheme, than before; the evil chiefly arising from want of efficient trapping and ventilation of the sewers."

Dr. Derby says: " That an obscure internal cause — which, in our ignorance of its nature, is called a proneness of disposition to receive the poison — is necessary for its development does not affect the truth of the fact that without filth the disease is not born. . . . . The improvement of public health as expressed by that unerring guide, the death-rate, corresponds with all the means by which air and water are kept free from pollution."

Typhoid fever is the most conspicuous type of the class of zymotic diseases, all of which are clearly pythogenic, and none of which can originate under conditions fit for proper human habitation.

A fertile soil or an impervious subsoil is especially favorable to typhoid poisoning; while deep gravel or sand, well drained, and offering free access to the air, are the least so. Rock near the surface is bad in the same way that a clay subsoil is bad, so far as the foundation of the house is concerned, while either rock or gravel may be, and often are, excellent. Neither is necessarily a certain security against the sanitary evils under discussion, for, through fissures in the rock and through the porous soil there is too often an ingress of polluted water from barn-yards, cess-pools, etc., or a spread of dampness from adjacent ponds or rivers.

It is doubtless too early in the education of the

community in this free country to expect the people to submit to the dictation of public authorities as to the manner in which they shall construct the foundation of their houses, but the time cannot be far distant when the public will assert its right to see that walls, foundations, and chimneys are so constructed as to preserve the public health — at least so far as to require that these important parts of home-building are made to conform to the health requirements of the situation.

As a rule, new residents in an unhealthy locality are more subject to disease than those who have become accustomed to the unfavorable influence; yet when typhoid contagion appears, it attacks first those whose systems have been debilitated by the insidious influences of foul air or water.

Malarial evils cannot be counteracted; they must be removed.

In 1874, an International Sanitary Congress was held at Vienna, at which it was unanimously affirmed, that there is no agent known which is certainly capable of destroying a contagion, and that we must look with suspicion, upon the efficacy of mere disinfectants.

A recent writer, discussing the epidemic of cholera, in Vienna, during the exhibition year, divides the causes of epidemic diseases, between *miasms* and *contagions;* the first being poisonous gases; and the second, germ cells. He states that contagion flourishes only where miasm is developed. Although this theory lacks scientific demonstration, it is in such close conformity with the results of all observation, that it may be taken as true in practical effect. Whatever may be the character or source of an infection, its development and activity are always fostered by the presence of decomposing matter, producing a miasmatic condition of the air.

Air, poisoned by stagnant water, in or on the soil, or corrupted by emanations from the decomposing wastes of human life, whether it originates an infection or not, is quite sure to aggravate any infection that may already exist.

Liebermeister says that typhoid fever " is not contagious in the proper sense of the word, for it is never transmitted by direct contact. It is not purely miasmatic, for external conditions alone are not sufficient to produce it. . . . . The development can take place if the dejections are left to themselves, as in dirty linen ; but it seems to go on more abundantly if the dejections are collected in privies, sewers, or ground already saturated with organic substances. In this way it can be explained how a typhoid patient who comes to a house or region previously free from the disease, can establish there a focus of infection from which many other persons become diseased. . . . . When typhoid fever is once established in any locality, it may disappear for a long time and then suddenly reappear, without the introduction of a new case."

Again he says : " From all that has been said, it results that the real cause, in our opinion, of every epidemic, and of every isolated case of typhoid fever is only the specific poison of typhoid fever. All the numerous conditions which have been called causes are not real causes. If the specific poison is absent, every other evil influence may act on the population without producing typhoid fever. No matter how well a field has been manured, wheat will not grow unless wheat has been sown." In like manner, the poison cannot be propagated unless the proper conditions are present, as with wheat, which, "if we sow on rocks, we sow in vain;" and so, " besides the presence of typhoid poison, many other conditions are necessary to produce typhoid fever." And again: " The cause of typhoid fever is always a specific poison; if this poison is not received into the body, anything else may be produced, but no typhoid fever."

One naturally argues from circumstances with which he is most familiar, and as I have given more especial attention to the sanitary short-comings of my own town, I take it as an example, believing however, that its interior arrangements are not less favorable than those of the average of our prosperous country places, and recognizing the important fact that its position (on a neck of land hardly a mile wide and sloping in one direction to the Atlantic Ocean and in the other to Narragansett Bay, without a hill or a forest to intercept the free-blowing winds from every quarter) makes Newport *naturally* a perfectly salubrious town. The population in 1870 was 12,521, the larger number living in a compactly built district facing the west and drained into a deep cove of Narragansett Bay. At the north and at the south the land is flat, but nearly all of it lies high enough for tolerable drainage. It is underlaid with stratified rock, and has a heavy clay subsoil interrupted by fissures of gravel sloping similarly to the surface of the ground.

There is no public water supply, and probably a large majority of the population drink water from wells only, although there are many filtering cisterns. Nearly all the houses of well-to-do people have the usual plumbing arrangements, which discharge into cisterns or into cess-pools in the ground. Some of these drain themselves through the gravel streaks of the subsoil, and a very few are absolutely tight, so that they require hand emptying. A rude sort of sewerage has been attempted here and there, laid without system and constructed apparently without the least reference to the well-known requirements of all town drains.

These sewers have the advantage of being at every opening so noisomely offensive that persons living near them are warned by the odor to keep their windows closed when the wind comes from a certain direction, and passers-by do not loiter in their vicinity. There is less insidious sewer poisoning here, as the exhalations are blazoned to the dullest sense. Usually where a sewer is available, the private cess-pool overflows into it, but in a great majority of cases the removal is by hand, with carts trundling into the country and making winter days and summer nights worse than hideous.

If the best winds of heaven did not blow almost constantly through our streets, we should probably be as badly off as a country town can be, but apparently this free ventilation will for some time continue to stave off the epidemic that awaits us, and which alone probably (here as elsewhere) will be able to secure the needed reform.

With these advantages and disadvantages Newport had a death-rate in 1863 of 34.16 per thousand (even supposing the population not to have increased between 1863 and 1870); a death-rate in 1873 of 25.76 per thousand, and an average death-rate for eleven years ending 1873 of 21.05 per thousand. The adjoining town (Middletown) which, like Newport, extends quite across the island, with a purely rural population of 1,074 persons, had in 1875 only 9 deaths, being a death-rate of only 8.38. Its *natural* sanitary conditions are by no means superior to those of Newport, — which owes its shamefully high death-rate only to the lack of intelligence with which it allows the accumulation of domestic filth to endanger the lives of its people and its visitors.

The town of Worthing, on the south coast of England, is probably more nearly like Newport in its climate, population, and uses than any other seacoast town with which it can be compared. Like Newport, Worthing is more or less a resort for invalids and persons seeking a beneficial change of air, but unlike Newport it has an excellent and abundant supply of pure water, and its drainage is said to be perhaps the most complete in Great Britain, every cess-pool and surface drain having been suppressed, and a main sewer emptying into the sea two miles away. The sanitary effect of this difference is shown by the fact that Worthing has the lowest death-rate ever recorded — 14.5 per thousand (during the second quarter of 1874 it was only 12.9 per thousand); and *a death-rate of* 14.5 *means an average longevity of nearly sixty-nine years for the whole population.* In 1874 there was only one death in Worthing from fever; this was certified as enteric (typhoid). It is probably as nearly certain as any such speculation can be, that could Newport have the simple advantage of a pure water supply and the perfect drainage that could so easily be given it, its average death-rate could be reduced to that of Worthing, causing us an annual saving of nearly one third of our deaths, with the enormous amount of costly and wearying illness and of debility and inefficiency that these deaths imply. Viewed in another light, could the questionable reputation under which Newport now suffers be replaced by one like that of Worthing, it would lead to such a growth of "stranger" population in summer and in winter as would gladden the hearts and overflow the coffers of all its eager army of purveyors.

All English watering places are not equally well cared for. In a series of articles describing the health of watering places in England we find the following statement which is specially recommended to the local governments of American sea-side resorts.

"For the benefit of those of our readers who may have little local knowledge of Whitby, it may be useful to refer to one or two matters which have a direct bearing upon the health of the town. First, as to the drainage. It appears almost incredible that in any place depending in great measure upon its reputation as a health resort, the whole of the sewage of a town of more than 13,000 inhabitants should be emptied into the harbor; and yet such is still the case with Whitby. The drawback to the enjoyment of Whitby, as a watering place, to say nothing of its health, arising from the condition of the harbor, especially at low water, has attracted considerable attention among the visitors during the season just ended, as we can aver from personal experience."

In fifteen other watering places described, the rate of mortality during 1874, was lower in twelve of them than in Whitby which was only exceeded by Southport, Falmouth, and Rhyl.

The degree to which the sanitary question has taken hold of the popular mind in England is well illustrated by the following, from the "Sanitary Record:" "Taunton has been for a long time considered to be — by its inhabitants, at least, if not by the world at large — 'the cleanest town in England;' and Tauntonians are accordingly just now greatly concerned on account of the fact that two persons in the town have died of typhoid fever; the victims being policemen, and the active cause of the fever being the presence in the town of a filthy slaughter-house which has not yet secured a sufficient amount of attention on the part of the board of health. The offending slaughter-house keepers appeared last week before the Taunton Bench, on an adjourned summons, tardily taken out at the instance of the local board of health. These cases of fever have attracted an unusual amount of attention, and have put Tauntonians 'quite in a flutter,' because the reputation of their town for general salubrity and exemption from fevers has hitherto been untarnished." When shall we see the day when two deaths from typhoid in a large town in America will put its people 'quite in a flutter?'

Nor are our cities better provided with sanitary appliances than our smaller towns. Even Boston, which congratulates itself on its refinement and civilization, is assiduously planting the seeds of future trouble.

The newer parts of the city, especially the district toward the Mill-dam, may serve as a very good illustration of what it is possible to do in the way of providing unfit habitations. The annoyances caused by the imperfect sewerage of this district have long been a ground of complaint even among persons who would accept the ordinarily defective drainage of higher-lying town-districts as quite satisfactory.

In this case the remedy though radical is simple, and it would be much less costly than would be supposed by those who are not acquainted with the artificial pumping system largely in use in England

SANITARY RELATIONS OF DRAINAGE. 65

Indeed, this district is especially well adapted for such drainage, for the reason that all its houses are occupied by a class who use water very liberally, so that there need be no fear that there would not be an ample flow to remove all solid matter reaching properly made drains.

All street-wash and the rain-water falling on the roofs, court-yards, etc. (beyond what would be needed for occasional flushing of the house sewers), may be removed by surface gutters or by a system of shallow drains discharging into Massachusetts Bay, and flushed, whenever needed, by water admitted to a flushing reservoir from Charles River at high tide. The house drainage itself should be disposed of through an independent system of small sewers laid at least three feet below the level of the lowest cellars, collected at one point and lifted by steam power into the sewer leading to Massachusetts Bay. Nothing but the fact that it is surrounded by wide stretches of water and great areas of unoccupied land could account for the preservation of the city in a state of even tolerable healthfulness, in the face of the circumstance that the water system is only partially introduced, and that one half of its night-soil, or about five thousand cords per annum, is still removed by carts : and it should be borne in mind that this five thousand cords is only what has been retained in the vaults after enormous volumes of its liquid parts have soaked away into a soil covered with a dense population.

Reference has been made to the fact that there is often less danger from impure well-water than from impure air, and some of the Massachusetts investigations indicate that in that State contaminated wells are not a very prominent source of infection. At the same time, the fouling of well-water by organic impurities is a very frequent source of fatal disease, and probably the reason why it does not appear relatively more serious in Massachusetts is that so much of the soil of that State is of a light character to a very great depth, there being less lateral movement of excessive soil-moisture than where strata of hardpan, or impervious soil, and seams in stratified rocks are more prevalent.

The reason why well-water is often bad and unwholesome is, in plain English, because sink-drains and vaults empty their foul contents into it. A well may be good for a long time and subsequently become poisoned, because the soil lying between the source of the impurity and the well itself has a certain amount of cleansing power. While this is effective, every impurity is withheld, but by degrees the soil becomes foul farther and farther on, until at last there is no grain of uncorrupted earth to stand between the sink and our only source of the pure water without which we cannot live in health.

The well is in effect a deeper drain, toward which the water from the surrounding earth finds its way, and in time, as impurities will follow water to any outlet unless the filter that holds them back remains always active, the foulness of the earth within the drawing range of the well is carried into the water, which it renders unfit for human use.

In 1847 typhoid fever broke out nearly at once in the thirteen houses of a single terrace in Clifton, England, which took their drinking-water from a certain well. Other houses in the same terrace escaped entirely, and it was found on investigation that in every house supplied from the well in question the disease was severe, while in no other house of the terrace did it appear. The infected houses were considerably scattered, and the only connecting link between the inmates was the use of the same drinking-water.

A very striking case in point which occurred in Williamstown, Massachusetts, was well and skillfully investigated. A house-drain became choked, and its contents mingled with those of a field-drain that was near a well. The season was wet, the ground was thoroughly saturated, and the surface water oozed into the well. The house was a boarding-house, with from thirty to thirty-five persons, mostly students, at table. Within two weeks most of the boarders were affected, and twenty or more of the students fell sick. At this time there was one case of typhoid fever in town, and this patient had been removed from his lodgings in the college to this boarding-house, where, probably by means of the escape of his dejections from the imperfect drain, his disease was communicated through the water of the well to all or nearly all of those who drank the water unboiled. Those who drank no cold water escaped: but a family in an adjoining house supplied from the same well were attacked, except one member who habitually drank no cold water. All who drank that water unboiled had the disease; all who avoided it in that state escaped.

Dr. Stephen Smith describes a visit to a country clergyman, a former schoolmate, who told him of a family, five of the members of which had died, while another was then fatally sick with typhoid fever; and he had not thought of attributing it to anything else but a " visitation of Providence." An investigation showed that during a busy harvest the valve of the pump had got out of order, and there being no time to replace it, water had been taken from a brook which received, higher up, surface water and the drainage from several barnyards. Of the entire family but two escaped attack, and they had not used the water.

The Broad Street pump in London is now famous in the annals of epidemics. During the cholera visitation in 1848–49, it killed five hundred persons in a single week. And many of the better classes, who fled the town and went to reside five miles farther up the Thames, were there attacked with cholera, it being found that they had been in the habit of sending to the Broad Street pump for their tea-water.

Having been instrumental in introducing the dry-earth system of sewerage into this country, it is proper for me to say here that my faith in the abil-

SANITARY RELATIONS OF DRAINAGE. 69

ity of that system to accomplish all that it has ever promised remains unabated, and that, under circumstances where its application is practically feasible, I should never recommend water sewerage; yet, in the present state of its development, it is so inapplicable to a large majority of cases, or so distasteful to a mass of persons whose necessities demand immediate relief, that, without in any way receding from its advocacy (to which a later chapter of this book is devoted), I freely acknowledge that the practical good which is to come of early sanitary reform is to be sought through other means.

The drawback, so far as towns are concerned, lies in the inability of this system to deal satisfactorily with copious amounts of water. Twenty-five gallons of waste running from a kitchen sink would require for its absorption from four hundred to five hundred pounds of earth. Still, earth sewerage can be perfectly depended on in village and rural establishments where there is a sufficient amount of lawn or garden to absorb the waste by underground irrigation; such irrigation beginning at a point sufficiently far from the house or the well. Disposed of in this way, and made to feed a vigorous vegetation, all of the liquid waters of the house may be safely treated in a small lawn or garden.

The evidence as to the sanitary completeness of this system is all as conclusive as the following recent report from a very unhealthy quarter: Before 1868, dysentery and fever were very prevalent in the convict-prison of Labuan, Borneo, and the old system of water-closets and cess-pits was abolished, earth-closets being substituted. Hereupon the sickness and mortality declined. In 1870 a great mortality broke out among the troops of the station, and a government inquiry developed the fact that in the barracks, where the earth system had been neglected, thirty per cent. of the troops died per annum; the deaths in the prison, where it had been assiduously used, amounted to but two per cent. In Sierra Leone, where the commanding officer had taken efficient measures to provide earth-closets for the troops, the health of the officers and men was maintained "at the very time when fever and dysentery were carrying off twenty per cent. per annum of the European population residing in the town."

A novel system of sewerage by pneumatic action, invented a few years ago by a Dutch engineer named Liernür, has been introduced on a large scale in parts of Amsterdam, Leyden, and other towns of Holland, and is now being much discussed by those interested in sanitary matters in England. The accounts given of the success of this method, of the entire absence of odor in all its processes, and of the complete saving for agricultural use of almost every part of the household waste, indicate that it is most efficient, economical, and admirable. The Pneumatic System is treated more fully in subsequent chapters.

CHAPTER II.

THE DRAINAGE OF HOUSES.

"The house is the unit of sanitary administration."

WHATEVER means may be adopted by the board of health of town or village for the removal of the wastes incident to the life of its population; whatever facilities for such removal may be offered by the natural surroundings of isolated country houses; and whatever the public or the individual may do toward rendering the natural character of the ground dry and salubrious, the first aim of the householder himself should be to secure a perfect means for carrying safely beyond the walls of his domicile everything of a dangerous character that is generated or produced within it, and to secure his living-rooms against the entrance of any manner of foul air, impure water, or excessive dampness.

It would not be possible here to consider the very great variety of circumstances attending the location and arrangement of different houses. It will suffice for our general purposes to assume that all liquid or semi-liquid drainage from the house is to be delivered either into a public sewer, into a private place of deposit, or directly into a natural water-course. If we arrange a safe means for discharging our outflows at a sufficient distance into one or other of these, for the exclusion from the house of gas arising from their decomposition, for preventing filtration from them into the source of our domestic water and for excluding soil-dampness, we shall, so far as exterior surroundings are concerned, accomplish the most important aim. General rules and principles, of which the modifications needed in particular cases will quite naturally suggest themselves, are all that can here be given.

The individual householder has these problems to solve : —

1. To secure his house against excessive damp in its walls, in its cellar, and, where practicable, in its surrounding atmosphere.

2. To provide for the perfect and instant removal of all manner of fluid or semi-fluid organic wastes.

3. To provide a sufficient supply of pure water for domestic use.

4. To guard against the evils arising from the decomposition of organic matter in or under the house.

5. To remove all sources of offense and danger which may affect the atmosphere about the house.

6. (And almost more important than all the rest.) To prevent the insidious entrance into the house, through communications with the sewer, cesspool, or vault, of poisonous gases resulting from the decomposition of the refuse of his own household,

or of other households with which a common sewer or drain may bring him into communication.

The first item implies a dry cellar, an impervious foundation wall, and, if the soil be heavy and liable to be wet, or if it be underlaid too closely with rock or clay, " thorough drainage," of the sort employed in the agricultural improvement of land. So far as this matter of drainage is concerned, it will suffice to refer to the well-known works on agricultural drainage; but the drying of the cellar and foundation receives so little attention at the hands of both owners and architects, that explicit directions seem advisable. If the house is founded on well-drained gravel or on a dry bed of sand (which is the best of all foundations) no further draining will be necessary; but even here it is always advisable to cover the floor of the cellar with an impervious concrete, to prevent the exhalation of moisture that arises from even the dryest soil; and in all cases where the foundation wall is not built with hard and impervious stone, it should be furnished with a course of some impervious material, whether hydraulic cement, asphalted brick, bluestone, slate laid in cement, or sheet-lead. An excellent asphalt for an impervious course in the foundation of houses is made of two parts of coal tar and one part of pitch with three handfuls of quicklime to each bucketful. Even with this precaution, if the foundation wall below the impervious course is of brick or soft stone, the inner surface of the wall should be well washed with pure hydraulic cement, which will lessen the escape of the moisture that penetrates the stones during driving rain-storms, or soaks into them from the earth.

If the ground is at all inclined, even in the wettest seasons, to be wet or springy, whatever other precautions are taken, a drain should be laid all round the cellar inside of the wall, and at least a foot lower than its lowest bed-stone, and carried away to a free and sufficient outlet. This drain may be made of gravel or broken stones, but ordinary land-drainage tile with open joints is usually cheaper and always better, especially as preventing the ingress of vermin. For the largest private house, the smallest-sized land-drain tile will be sufficient. If the soil is unduly wet, at any season, similar drains should cross the cellar at intervals of not more than fifteen feet. All of these drains should have a slight but continuous fall toward the outlet, and should be securely covered by having earth well rammed over them, the whole cellar bottom being then coated with concrete. For small houses, where cobble-stones or gravel are plenty, if the foundation rests on a layer of this porous material a foot or more deep, and if a good outlet be provided at the lowest point, the tile is not needful.

The complete drainage of the house, that is, the instant removal of the impurities incident to human life, is the crowning work of the whole system of sewerage. In towns, street drains, main sewers, outlets, and the whole paraphernalia of the system have for their main purpose the furtherance of the ultimate object of the sanitary drainage of the house; and the effect of sewerage on the health and decency of the population must depend very much upon the manner in which each house is provided with the necessary drainage system and is connected with the public sewer.

In the country, whatever the final means of removal, the house drainage, whether partial or complete, must be equally guarded. If there is only a kitchen drain, this should be perfectly well made and arranged.

When we consider its immediate proximity to the windows of the room in which the family of the average farmer passes most of its time, the kitchen drain probably heads the list of all the agents by which our ingenious people violate the universal sanitary law; and it doubtless carries more victims to the grave than do all other sources of defilement combined; for with an enormous majority of our population this one pipe still represents the whole drainage of the house.

Receiving daily supplies of organic matter ready to pass into dangerous decomposition, drenched with sufficient water to soak far into the ground, and kept warm enough for putrefaction to proceed rapidly throughout a large part of the year; sending its exhalations into the kitchen and living-room windows, and with a favorable summer breeze throughout the whole house; and leaking, too often, through a light surface soil, or through a porous stratum in a clay soil, into the adjacent well; it attacks the family through the lungs and through the stomach with an almost unremitted assault, soon achieving, in the case of those who live mainly in close, stove-heated rooms and sleep on the ground floor with a window opening over the back-yard, its various measures of debility, disease, or death.

No house drain can be made which may not be carelessly obstructed by the admission of substances for which it is not intended. Shedd enumerates, among the articles that have been found in such drains, " sand, shavings, sticks, coal, bones, garbage, bottles, spoons, knives, forks, apples, potatoes, hay, shirts, towels, stockings, floor-cloths, broken crockery," etc.

House drains in towns should of course be laid at the expense of the owner; but, as they are a part of the system by which the health of the community is to be protected, and as the obstruction of a single house drain may establish a centre of infection for a large district, the work should be done in accordance with an established rule and under the immediate supervision of an engineer having charge of sewerage work.

The main outlet drain from a house may be small, and even for the largest private dwelling need not be more than four inches in diameter, if proper precaution is taken to prevent its being choked by the accumulation of kitchen grease; while, without such precaution, were it even a foot in diameter, this same influence would cause it to be ultimately obstructed by gradual accumulation. In other words, with a proper grease trap, a four-inch drain will furnish an ample outlet, while without such grease trap no drain can be relied upon to remain permanently effective.

There are various forms of grease trap, some with open gullies for ventilation at the surface of the ground, and all of them depending upon the congealing of the grease and its accumulation at the surface of water which has its outlet at a point below the surface.

The removal of organic wastes is the chief purpose of all house drains, whether a wooden pipe from the kitchen sink, or the soil pipe of a house fitted with all the modern plumbing appliances. It is this part of the work that first suggests itself when the question of house drainage arises, and it is too often to this only that attention is given.

The water-closet, owing to its convenience and seeming cleanliness, has made its way to almost universal adoption, in spite of a very serious defect which is generally disregarded, and, indeed, unrecognized. This defect consists in the use of an unventilated chamber between the sealing-pan and the water trap of the soil pipe,—a chamber that is always more or less foul, and where fæcal gases are constantly generated. No means are provided, and no perfect means could be provided, for the removal of these gases, which are sure to find their way more or less into the atmosphere of the house, if only by transmission through the water seal. Persons living in the country think that they can always detect the odor of the closet on entering a city house, and this odor is very often due to the cause here indicated. It is only very recently that inventions (the Jennings closet, and later, Smith's closet) have been made which entirely overcome this defect; although several other forms of closet using large volumes of water and not depending upon a tilting pan for their sealing seem to escape it.

Some of the minor devices of modern plumbing seem as objectionable as they are convenient: for example, the ordinary fixed wash-basin having a plug at its bottom effects a complete separation between the water in the basin and the foul, soap-slimed escape pipe below it; but the more convenient shut-off cock placed some distance below the basin is a most ingenious arrangement for tainting the water in the basin, which is in free communication with the water in the unclean escape pipe. How readily impurities are diffused through still water is shown by the rapid clouding of the contents of a tumbler to which a used tooth-brush has been returned; the invisible solution from an unclean waste pipe spreads with equal ease.

It is now quite usual, also, to ventilate the lower chamber of the ordinary water-closet, and this is to a certain extent effective for the purpose intended; but it does not accomplish a proper ventilation of the soil pipe, and it alone should by no means be depended on. Indeed, this lower chamber is always objectionable, sending forth such a whiff of fetid air, whenever the water pan is emptied, as could come only from a confined, dark, and wet vessel where the most offensive matters are undergoing decomposition. The cheap and simple siphon-closet, with a copious flow of water, or, better, the Jennings closet, with Blunt's overflow, or Smith's closet with heated ventilator, are types of the only satisfactory forms.

In the country and in villages, where each house has to be provided not only with the ordinary interior arrangements, but also with means for the disposal of its drainage after this has passed beyond its own walls, a serious further difficulty arises. The usual practice, where plumbing is introduced, and very often where only the water of the kitchen drain is to be provided for, is to discharge the whole mass into a cess-pool not very far away, and often very near to the well, trusting to the permeability of the earth to afford an outlet through the uncemented wall. The objections to this have been sufficiently stated, and the remedy is not in all cases an easy one.

There is no royal road of escape from the responsibility that the production of effete matters entails upon us. If they can be run by a cemented drain into a water-course, or elsewhere, far enough away from human habitations to be unobjectionable, this course may be allowed; but in the great majority of instances it is absolutely necessary to provide for their defœcation in some inoffensive manner or for their inoffensive removal by carts to the country. The one thing that should never be allowed in a village, and which should even be regarded with great caution in the case of an isolated house, is the ordinary leaching cess-pool.

The importance of a close attention to the details of household drainage cannot be overestimated. In a paper on the Health of the Farmers of Massachusetts, Dr. Adams of Pittsfield says: "There is no dwelling in the State, of any class, which possesses an absolute immunity from these causes" (the vicinity of putrescent animal and vegetable matter); "for they are often so hidden and subtile as to elude the search not only of the landlord, but also of the most vigilant health officer."

The securing of pure drinking-water for the household can hardly claim full attention here, except so far as drinking-water wells are concerned; although the extent to which water coming from public works is contaminated by an injudicious arrangement of supply pipes and soil pipes is often alarming, as was suggested in the preceding chapter.

Quite generally, country houses and houses in villages and towns depend entirely upon drinking-water wells for their supply, and the degree to which these wells are rendered dangerous by what is called "surface water"—that is, rain-water passing over or through a surface soil made foul by house slops, kitchen refuse, etc.—is more than alarming. The purity of the water in any well depends almost entirely on the ability of the earth through which it descends to deprive it, by filtration, of its organic impurities. It is always a question between the amount and character of the filtering material, and the amount and character of the impurity. In a deep, porous soil, where the water-table lies at a great depth, and where the rain-water descends vertically to the line of saturation below, there is little danger, unless the grossest fouling of the surface in the immediate vicinity is constant and long-continued; but where the water level is near the surface of the ground, where the soil has an impervious stratum a few feet below the surface, or where the well is supplied through rock fissures or gravel seams which open near to the surface of the earth, the most scrupulous cleanliness is needed to prevent contamination.

Fresh earth is a capital purifying filter, and the rapidity with which its filtering power is renewed depends upon the freedom with which air circulates within it, the purification being in nearly all cases a process of oxidation. In a deep and porous soil, as the water of a rain-storm settles away, it is immediately followed by the entrance of air from the surface, and the oxidation may be complete; but in clay and in other impervious soils, the entrance of air being much more slow and difficult, the impurities accumulate and the foulness increases and too often becomes permanent. In soil of this character the curbing of the well should be laid in cement for some distance below the surface, and wet clay should be closely puddled round the curbing, to prevent the trickling down of water between this and the solid earth. The greater the distance between the surface of the ground and the point at which water first oozes from the earth into the well, the greater the safety.

The above refers only to the fouling of wells by the leaching down of impurities thrown upon or accumulating in the surface soil. A much more frequent and much more serious source of mischief is their contamination by foul water leaking from badly made house drains or flowing laterally from cess-pools or vaults,— our own or our near or distant neighbors',— or the trickling down through gravel seams in heavy soils or porous fissures in rock of the surface liquid of adjacent or remote barn-yards. When porous strata in rock or earth bring the site of the cess-pool into communication with the site of the well, the danger is immediate and constant until the cess-pool is made absolutely tight.

The more insidious process is that of the gradual fouling of the semi-porous earth lying between the source of the impurity and the drinking-water well. In such cases the exudation is usually quite or nearly constant; there is little opportunity for the air to restore the filtering power of the soil, and it becomes saturated with impurity inch by inch, until, perhaps after a month or perhaps after several years, the saturation reaches the well; then every drop oozing in from this source carries with it its atom of filth. While the supply of water in the ground is copious, and while there is more or less circulation through the water veins, the foulness may be too much diluted to do harm; but in dry seasons, when the supply recedes to a depth of only a few feet at the bottom of the well, the contribution of drain water continuing the same, the dose becomes sufficient to produce its poisonous effect.

The dangerous character of the water of such wells, is often manifested by no odor or taste of organic matter; the chemical changes in this matter seem to have been carried so far, as to yield little more than vivifying nitrates to the water, their organic character having entirely disappeared. Indeed, some of the most dangerous well-waters, are especially sparkling and refreshing to the taste. But the chemical processes which have effected this change, appear to have had no effect on the germs of disease — if germs they be — which retain their injurious character to such a degree, that the worst results have often come from the use of water that was especially sparkling and pleasant as a beverage.

The bad effects of organic decomposition, are nowhere more manifest than when it takes place in an unventilated cellar.

That large part of the American people who were born and bred in the country, will appreciate the following quotation from Judge French, describing childhood's experiences with New England cellars: "You creep part way down the cellar-stairs, with only the light of a single tallow-candle, and behold by its dim glimmer an expanse of dark water boundless as the sea. On its surface, in dire confusion, float barrels and boxes, butter-firkins, wash-tubs, boards, planks, hoops and staves, without number, interspersed with apples, turnips, and cabbages, while half-drowned rats and mice, scrambling up the stair-way for dear life, drive you affrighted back to the kitchen." This will seem to many like exaggeration, but probably throughout all America, one half of the population which lives over cellars at all, lives over cellars which, at some time during the year, approach the condition described.

All this nastiness and wet and confined moldiness and stagnation, must inevitably foul the air of the whole house, rendering it impure to a degree that makes us wonder how human beings, if they can live at all, can live in even tolerable health in such abodes.

A medical correspondent of the Massachusetts Board of Health, gives an account of the cellar of a house in Hadley, built by a clergyman, which had an uncovered well within it, and into which a sink drain with its deposit-box had full opportunity to discharge its gases, there being no proper ventilation to the drain or box. The cellar was also used for the storage of vegetables, and its windows were never taken out. There was no escape for the damp and foul air of the cellar, save through the crevices of the floors into the rooms above.

"After a few months' residence in the house, the minister's wife died, of fever, so far as I can learn. He soon married again, and within one year of the death of the first wife the second died, from, as I understand, the same disease. His children were also sick. He lived in the house about two years. The next occupant was a man named B——. His wife was desperately sick. A physician then took the house. He married, and his wife died of the fever. Another physician was the next occupant, and he, within a few months, came near dying of erysipelas. All this time matters had remained as before described, with reference to ventilation." After this a school-teacher took the house, and made some unimportant changes. "The sickness and the fatality of the property became so marked, that the property became unsalable. When last sold, every sort of prediction was made as to the risk of occupancy, but by a thorough attention to sanitary conditions, no such risks have been encountered."

It is hardly necessary to recur to extreme instances of cellar foulness, such as those above described, to convince any person of ordinary intelligence that in a confined and dimly-lighted atmosphere, like that of an ordinary cellar, all decomposition of organic matter must result in the production of gases unfit for human breathing.

We especially need a condition of air, that can be maintained only under the influence of light and free ventilation. The great difficulty with our cellars is, that as they have a more or less complete communication with the house through open doors, imperfectly laid floors, etc., and as the law of the transfusion of gases is constantly operating even though the means of communication may be imperfect, their unceasing tendency day and night is to communicate their impurities to the air of the house. Where floors are deafened and where the ceiling of the cellar is lath-and-plastered, the danger is much less than where there is only a single thickness of boards with imperfect joints to check the communication; but no matter how perfect the separation may be, everything like the decomposition of vegetable or animal matter should be studiously avoided, and there should be at all times a free (though slight) circulation of air in the cellar.

To live in a house standing amid offensive and dangerous surroundings, is under no circumstances either necessary or excusable. It has been well said, that no man is so poor that he need have his pig-trough at his front door, or that he need throw his slops under his dining-room window. No place is so small that it need contain a fermenting manure heap, within dangerous proximity to the house; either the fermentation must be arrested, or the accumulation must be entirely avoided. No yard is so flat that the slops of the house may not be drained away to a sufficient distance for safety. In short, there are in this world no circumstances unfit for wholesome living, which may not be either overcome or run away from.

To live wrongly is a danger and a disgrace to the individual. To permit such wrong living is more than a danger and a disgrace to the community; it is a criminal menace to its own health and life. No special rules and regulations need be given here for the avoidance of palpable sources of danger; all that is necessary is studiously to avoid the retention of accumulations of organic filth of whatever description.

In the living of every family there is a certain necessary production of waste organic matter. In the economy of nature all such waste is destined to return to its elementary condition and to become a part of the air or soil or sea, awaiting its renewed use in feeding plant life. Man has learned how to avail himself of nature's organic products to supply his demand for food and clothing. He seems not yet to have learned how to hand back to the realm of nature the refuse that is not useful to him, in such a way as to avoid the injury with which its neglect threatens him. Were each man dependent only on the conditions in and about his own house, it would be safer than it now is to leave the needed reformation to individual action; but unfortunately all in the community are dependent for life and health more completely than they realize on the condition and surroundings of their poorest and most ignorant neighbor.

The public has long asserted and exercised its right to abate nuisances, but its definition of the term "nuisance" begins at a point far removed from the stricter sanitary limit. Our communities seem not yet to realize that they have a clear right to the health and strength of their individual members, and especially to insist that no man shall, by incurring the risk of disease in his own family, endanger others to whom his disease may be communicated. The stamping-out process must extend to the very bottom of society, and if we apply ourselves to the stamping out of causes, the effect (infectious disease) will not demand our attention.

If slops thrown out at the kitchen door of the poorest house in the town threaten the health of those living in that house, all who may eventually suffer from the sickness beginning in that family have as clear a right to prevent the cause as they would have to put the family in quarantine were they suffering from small-pox.

The art of purveying has been brought very near to perfection, and it may well be left to the commercial instincts of those whose business it is; but the hardly less important art of scavenging has received from the outset nothing but the sheerest neglect. It is as yet hardly in its infancy; if we can hide our filth away underground, shove it down the gutter to our neighbors' premises, or secrete it in one of those fermenting retorts, a public sewer (as usually constructed), we think we have done all that our own safety requires of us, and the safety of others we have not yet learned to regard. But our own safety is by no means secured; we are always in danger from our neglected wastes, and in proportion as we and others use the common sewer do they endanger us and we them.

That precursor of the sewer, the receptacle in our own yards, is certainly less dangerous than its modern substitute, but it is usually very far from being a safe neighbor to even an isolated house. As houses accumulate, the risk from it increases.

I was recently conversing with an intelligent builder about the construction of a contemplated cess-pool.

"It is useless to suppose that in such heavy ground a filtering cess-pool can very long answer a good purpose."

"I don't know how that is, but my own cess-pool in similar ground has been in constant use for eight years without being cleaned out, and it works all right yet."

"How do you know that it is not leaching into your well?"

"Because I put my well a good distance away from it, on the up-hill side."

"How about your neighbors' wells, down the hill below you?"

"Oh, I don't know anything about them; that's their lookout."

The fact is that the whole hill-side near the top of which this man lives is supplied with alternate cess-pools and wells, and there is every reason to suppose that the porous strata through which the cess-pools are emptied are the very strata from which the wells are filled.

It is not to be understood that the ordinary out-house vault is necessarily a source of danger; there is enough to be said against this arrangement on the score of convenience and of decency to serve as an argument for its abolition; but if it be cemented perfectly tight, and if its contents be daily disinfected with carbolic acid, sulphate of iron, or other suitable chemical addition, there is no fear of either the poisoning or the dangerous fouling of the air.

So, too, if it be used only for its legitimate purposes, if no liquid matter of any sort be poured into it, and if it have a copious daily sprinkling of ashes or dry earth, it will be equally free from sanitary objection.

But if these precautions are not adopted and carried into effect under a rigid supervision, there is certainly no single appurtenance of the life of an ordinary household so fraught with danger and annoyance to all who live within reach of its influence.

In all towns and villages where this expedient is allowed to remain in use, the strongest and most persistent effort of the health authorities, reinforced with full power for the infliction of penalties, should be devoted to the regular, frequent, and efficient supervision and inspection of every out-of-door closet of whatever description. The removal of the contents should never be left to the uncontrolled action of the proprietor, but should be carried out by well-directed workmen in the employment of the board, or at least under its direct inspection.

However perfect the ventilation of sewers or cesspools, the safety of individual families and of all to whom they may communicate disease demands a careful attention to the ventilation of the house drain. It is chiefly through this drain that cess-pool and sewer gas finds its way into the house, and the house drain itself will be relatively more foul than will the public sewer which takes also the wash of streets. Dust and foul matters of various sorts are very apt to accumulate with the congealed grease that so frequently coats the sides of the drain. Therefore, so far as the house itself is concerned, its greatest source of danger lies in the return to its rooms of the emanations from its own offscourings, entering through the water traps or through leaks in the pipes, whether such return be caused by their own expansive force or by the pressure of the sewer air behind them. Chemical disinfection can, at best, afford only temporary relief. Dr. Simon says on the subject of the disinfection of houses which are said to have offensive smells or which inspectors find in a stinking state, " It cannot be too distinctly understood that *cleanliness* and *ventilation* and *dryness*, are the proper deodorizers of houses, and that artificial deodorizers will no more serve in their stead than, in regard of perfumes, these could serve instead of soap and water."

The following extracts are taken from Dr. Robert Angus Smith's work on " Disinfectants and Disinfection."

"Animal matter, which chiefly is found to be

dangerous, is, in fact, the fæces or *dejecta* of human beings and of cattle. It might be supposed that these substances had already been decomposed, but such is not the case. The decomposition is very imperfect, and when they are allowed to stand putrefaction sets in, closely allied to, perhaps exactly the same, as that which takes place in other animal matters, such as blood, or in a mixture of flesh and water. When these substances decompose, the result is, so far as we know, nearly the same as the decomposition of the entire animal body. We are not able to tell the difference between the products of putrefaction from our cess-pools and those from our graveyards. The problem, then, of preserving meat, or of preserving the entire animal from corruption, and the problem of preserving sewage and fæces from decomposition, become entirely one and the same. We are required to do for the fæces that which the Egyptians did for their bodies, until they shall be thrown upon the ground, and mixed with the soil and become the food of plants. . . . .

" Every substance in fine powder disinfects — dust of all kinds, whether platinum powder or powder of sandstone. The surface is enormously increased in such bodies, and surfaces attract the air, which is confined and pressed into service, causing mere active oxidation, and therefore more purification. . . . .

" Pettenkofer says that carbolic acid preserves inert the ferment cells, but when it is removed they become active. If this is true, the disinfectant must be used continuously, and the impure matter must be cleared away continuously, whilst soon in time, *and especially in the earth* the infectious matter will die. . . . .

"One may very correctly look on the soil as the greatest agent for purifying and disinfecting. Every impurity is thrown on it in abundance, and yet it is pure, and the breathing of air having the odor of the soil has, on what exact evidence I do not know, but very generally, been considered wholesome."

It is by no means enough to establish even the most perfect water trap in the line of a house drain. It is of at least equal importance that there should be a free vent for the escape of all air from the sewer and all gases generated within the house drain or in the soil pipe, not into the attic of the house nor at its eaves, near sleeping-room windows, but well up through and above the highest point of the roof.

House-drain ventilators are often introduced into chimneys, but they are nearly as often removed after a short trial. So long as there is a constant upward draft in the chimney, this disposition of the gases is good enough, but when no fires are used, the chimney frequently becomes a down-cast shaft, or when gusts of wind drive the smoke or the soot-smelling air into rooms, the ventilator gas is sure to accompany it.

House drains are even more liable to changes of temperature, and therefore more subject to a varying pressure of the air within them, than are sewers of themselves.

What is known under the general term "sewer gas" is the emanation from waste matters undergoing decomposition in the absence of free air and light, and in the presence of water, whether in a sewer, a house drain, a cess-pool, a vault, or a foul, wet, and unventilated cellar. It frequently exists in the case of a detached country house, and is never absent from a town sewer, though it is possible in the case of these, by perfect ventilation, greatly to lessen its production, and so to dilute it as to prevent its doing serious harm.

Poisonous sewer gas cannot be clearly defined. It is known chiefly by its effect; even its odor is rarely a marked one, and danger is believed to lurk not so much in those foul stenches which appeal to our senses, as in the odorless, mawkish exhalations which first announce themselves by headache and debility. This gas, in its most dangerous form, is believed to be some product of organic matter undergoing decomposition in the presence of superabundant water, and in the absence of light and free ventilation.

It may be present without detection; and, in addition to its frequent passing of the usual water traps, it is largely drawn into our living-rooms by the draught of heated chimneys when their demand for air is not abundantly supplied through other and easier channels of ingress.

Furthermore, soil pipes, as frequently constructed, crack or open their joints, by the frequent expansion and contraction that alternate floods of hot and cold water occasion, and thus give vent to their gases.

It is well known that leaden waste pipes decay and frequently become so perforated as to allow the escape of gas into the house. This decay always takes place from the inside, and generally at the upper sides of a pipe running in a horizontal or oblique direction; that is, in the horizontal pipe leading from a closet to the descending soil pipe more often than in the descending soil pipe itself. If there is a bend in the pipe the perforation occurs in its upper part. It usually occurs first in the highest pipes in the house. The perforation is very much the most rapid in the entire absence of ventilation. If the ventilation is by means of free and clear openings above and below, the corrosion amounts to very little.

The fact that the point of attack lies in that part of the pipe which is not covered with water, and more especially in the higher portions, — to which heated sewer gas at once rises, and where it accumulates, — indicates clearly that the corrosive action is due to the resultant gases of fæcal decomposition and not to the liquid contents of the pipe. As the corrosion begins on the inside of the pipe, and at a point where perforation would not ordinarily cause leakage, it is very likely to be overlooked, even when sought for by a plumber appyling the usual tests for leakage.

The diseases resulting to the inmates of the house from this condition of the pipes, and from other means for the admission of sewer gas, are those usually caused by excrementitious poisoning: namely, typhoid fever, diphtheria, diarrhœa, cerebro-spinal meningitis, scarlet fever, etc.; and it is always safe to advise any one in whose house such diseases appear, to uncover his soil pipes and have them thoroughly examined. Dr. Fergus, in his pamphlet " The Sewage Question : with special reference to traps and pipes " (Glasgow, 1874), says : " Lead has generally been used as the material for soil pipes, and as we have seen how capable it is of corrosion, it becomes a very important sanitary question to inquire how long a good lead soil pipe will hold out. I have been studying this question for years, and it is now about seven years since I first exhibited decayed pipes in public, yet, I would not wish to dogmatize on the subject, but rather give approximations, and would remark that the time will vary under the various circumstances according to the strength and rapidity of the flow of the sewage, as well as the original thickness of the pipe. But after allowing for this, we must broadly distinguish between soil pipes which are ventilated and those which are not. By the former I mean when the pipe is carried up to the roof of the house and open to the external air; by the latter, I mean when the pipes are closed up. Of these last mentioned, the duration may be stated to be about twelve years, the extremes of variation being from a minimum of eight to a maximum of twenty years. In ventilated pipes the duration may be stated to be nearly double,

THE DRAINAGE OF HOUSES.   97

running from twenty-one to thirty-three years, the extremes of variation being from eighteen to thirty or even more years. The practical sanitary conclusion which it concerns us all to keep in mind is, that any house, no matter how carefully or well built, may become unhealthy from this source, and that when cases of typhoid fever, diphtheria, etc., occur, the pipes should be thoroughly inspected, especially their upper surface, and the whole of the soil pipe uncovered. I must strongly insist on this, as in many cases the plumbers have declared pipes to be all right, which turned out to be very defective when uncovered. For some years back, I have insisted on a careful examination of the soil pipes wherever I have cases of typhoid or diphtheria, and in every case where I could get this carefully carried out, I have detected these perforated pipes, or sewer air getting into the houses in some other way."

One of the other ways he believes to be by the transfusion of the gas through the water of the trap, which he seems clearly to have detected. In experiments with glass pipes having bends or water traps it was found that the light gases passed through by the top of the bend, and the heavy gases by the bottom. "The action of the gas was curious. It was found, first, to saturate the surface of the water next to it in the trap; then to sink down in a fine stream, and then gradually travel through to the other or house side of the trap, when it again spread out and began to diffuse itself both into the atmosphere above it and downward through the water also." More recently Dr. Fergus made a series of experiments with a bent tube, the bend being filled with water after the manner of the usual trap. In the sewer end of the tube he inserted a small vessel containing a solution of ammonia. In fifteen minutes the ammonia had passed through the water of the trap, and had bleached the colored litmus paper exposed at the house end. In another experiment he produced the rapid corrosion of a metal wire exposed at the house end. To prove that this transmission takes place not only with ammonia, which is lighter than air, he made the same experiment with sulphurous acid, sulphuretted hydrogen, chlorine, and carbonic acid, all of which were transmitted so as to produce their chemical effect on the other side of the trap in from one to four hours.

Dr. Carpenter of Croydon, England, says, " The demands of air for fires and for respiration must be supplied from some source, and as the easiest means of access is often the communication between the house and the sewer, the poisonous gases which are lightest, and therefore in the highest parts of the drains nearest at hand, are first drawn in."

He gives the following means by which the admission of these gases is obtained: through the water-closet trap, when the soil pipe itself is unventilated; through defective joints or fissures in the soil pipe, resulting from bad workmanship, accident, or decay; through the waste pipes of house-maids' sinks, butlers' sinks, kitchen sinks, and untrapped bath outlets; through the overflow pipe from wash-basins, etc. He especially emphasizes the emptying of the traps by siphon-like suction, or, where the trap is not constantly used, by the evaporation of the sealing water. He thinks that not one trap in ten thousand is properly protected, and that without protection they are worse than useless.

The healthful arrangement of the water supply and drainage of the house in its minutest details ought to be a chief care of the architect, whereas, in practice, it is almost invariably left to a plumber, doing the work too often by contract, and having no conception of what is needed, — only of what has hitherto been done.

Evils resulting from the admission of sewer gas into living-rooms are popularly called "accidents," but they are accidents which may always be foreknown and the prevention of which is perfectly understood; they are no longer accidents, but gross faults of commission.

Until lately, in applying the water system, it has been considered sufficient to interpose what is called a water trap — usually an inverted siphon, in which water is supposed to be always standing — between the house and the waste pipe leading to the sewer. These traps, as commonly constructed, are in every way defective: even a light wind blowing into the mouth of the sewer often increases the pressure sufficiently to send the sewer gas bubbling through them into the house; a great rush of water into the sewer during heavy rains, by lessening the air-space and compressing the contained air, often similarly forces the traps. The same effect is produced in a marked degree by the rise of the tides into the mouths of outlet sewers in sea-side towns, the air being compressed into a smaller space and forced to find a vent. Even did these difficulties not exist, the fact that water transmits aeriform matter would always remain as an objection ; sewer gas is absorbed by the water of the end of the trap which is toward the sewer, and is given off to contaminate the air at the end which has a communication with the interior of the house.

The ordinary soil pipe has a trap at its lower part, and, if unventilated, its air stagnates often for hours together. When the pipe is used, every gallon of water poured down causes a displacement of some of the contained gas, which will seek its easiest means of escape, probably into the house. When a current is set up in a siphon-shaped trap below, its contents necessarily vibrate back and forth for a certain time, giving an impulsion to the confined air above that will tend to force it through fissures or feeble traps.

All soil pipes should deliver into ventilated traps outside of the house so that gases forced from the sewer need not even depend upon the open ventilator from the top of the soil pipe, but shall not be allowed to escape into the house at all.

Instances of fatal disease arising from imperfect plumbing have been and still are numberless, many of them as glaring as the following, described by Hakerman, who says that in the prisons at Brest the apartments which were supplied with water-closets were filled with sewer gas when the southwest wind drove the air through the sewers and forced the traps. In these apartments the cholera raged with great intensity, while those parts of the prison not supplied with closets remained free from it.

Dr. Fergus asserts that diarrhœa, cholera, diphtheria, and dysentery have increased fourfold since the introduction of the water-closet system with its numerous inlets for sewer gas into houses and water-supply cisterns.

In 1872, the Medical Officer for Edinburgh reported that wherever water-closets were introduced, "in the course of one year there were double the number of deaths from typhoid and scarlet fever, and any epidemic fever occurring in these houses assumed a character of malignant mortality." In our own cities it is known that the fatal prevalence of typhoid, and it is believed that frequent epidemics of diphtheria and cerebro-spinal meningitis, are due to faulty drainage alone.

In doing away with cess-pools and substituting sewers, unless proper precautions are taken, we simply make an elongated cess-pool, rarely sufficiently cleansed, and often grossly foul, and communicating with the interior of every dwelling-house. If typhoid excreta are thrown into a sewer a mile away from us, we have no security against the danger that its poisonous contagium will float in the gas of the sewer, and enter our own living-rooms. Grave as this difficulty is, it may be almost entirely removed by a proper arrangement of the drainage works of the house itself.

How slight a change in temperature in a sewer or cess-pool suffices to force a water trap, may be seen by experimenting with the simple apparatus illustrated herewith.

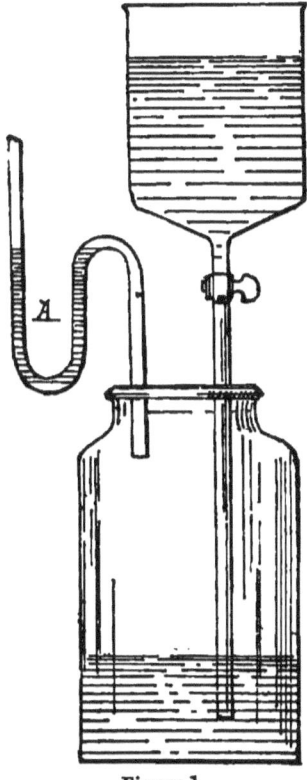

Figure 1.

The bend in the pipe A, filled with water, represents the common trap of house plumbing; the flask is filled with air. If the hand be simply placed upon the flask, the bodily heat will expand the air sufficiently to throw the water quite out of the pipe, although its upper arm may be several inches long. In like manner, on opening the cock in the pipe leading from the vessel above, containing water, the contents of this vessel will flow into the jar and bring to bear upon its contained air such a pressure as will force the water out of the bent tube. This represents precisely the condition of things when the quantity of water in the sewer is materially increased by sudden rains or by the rise of the tide into the outlet.

Another cause of changing pressure upon the air

THE DRAINAGE OF HOUSES. 103

of the sewer, is the frequent ebb and flow of the volume of sewage, now only a thread of water along the floor, and now an amount sufficient to fill it to half its height.

The ventilation of soil pipes is not only needful to carry away sewer gas, which would otherwise be forced through the traps or transmitted by their water, but also to prevent the formation of a vacuum when large volumes of water are poured down them. The vacuum thus formed is quite sure to suck open one or more of the water-traps, — which, until it is filled at its next use, will remain free for the passage of the gas from the pipe into the house.

A soil pipe in untrapped communication with a sewer, has been described by Dr. Carpenter, as an elongated bell-glass, affording a certain depot for the lighter products of decomposition and putrefaction; if the soil pipe has a free ventilation by a direct channel to the outer air above, these gases will escape harmlessly, but unless such outlet is provided, they will themselves seek out (or create) defective spots through which to find their way to the interior of the house.

Unused water-closets are especially dangerous, as the water in the trap, which was their only feeble barrier to the communication between the inside of the sewer and the inside of the house, is soon removed by evaporation ; and as ordinarily arranged, the overflow pipes of little used bath-tubs and stationary wash-basins, have their traps empty and open during a large part of the time.

In the very complete sewerage work of Croydon, Dr. Carpenter early insisted upon the compulsory ventilation of soil pipes, but his opinion and advice had to be reinforced by a long list of deaths traceable to the lack of ventilation, before the authorities adopted the rule. The work was systematically carried out by Mr. Latham, who was then a director of the Croydon board, and who has since become a leading authority in matters of sanitary engineering. Although he had himself given full credence to Dr. Carpenter's belief, he was astonished at the result. " Typhoid was sprinkled here and there before him; but as the work progressed it entirely disappeared from behind him and has not returned." Since this statement was made, typhoid has occasionally reappeared in Croydon, — owing to the fact that preventive regulations were in some cases made applicable only to *new* houses, defective arrangements already existing being allowed to remain. A recent severe attack has led to more complete reforms, and it is hoped that a better condition as to health will result.

Popular attention is now being resolutely drawn to these important sanitary considerations, and we may reasonably hope that we have fairly entered on an era, in which the improvement of sanitary conditions will be an important attendant of advancing civilization.

In a later chapter of this book, more explicit practical directions concerning the drainage and ventilation of houses, will be given.

CHAPTER III.

THE DRAINAGE OF TOWNS.

"All filth is absolute poison." — BOWDITCH.

IT should be the first purpose of town sewerage to remove the unclean refuse of life rapidly beyond the limit of danger; the second, to prevent it from doing harm during its passage; and the third, to regulate its final disposal.

The channel through which the removal is effected — the sewer — whether large or small, must conform to certain conditions, or it had better never have been built: —

a. It must be perfectly tight from one end to the other, so that all matters entering it shall securely be carried to its outlet, not a particle of impurity leaking through into the soil.

b. It must have a continuous fall from the head to the outlet, in order that its contents may "keep moving," there being no halting to putrefy by the way, and no depositing of silt that would endanger the channel.

c. It must be perfectly ventilated, so that the injurious gases that necessarily arise from the decomposition of matters carried along in water, or adhering to the sides of the conduit, shall be diluted with fresh air, and shall have such means of escape as will prevent them from forcing their way into houses through the traps of house drains.

d. It must be provided with means for inspection, and, where necessary, for flushing.

e. Its size and form must be so adjusted to its work, or to its flushing appliances, that the usual dry-weather flow may be made to keep it free from silt and organic deposits.

A sewer that is deficient in *any one* of these particulars is an unsafe neighbor to any inhabited house, and a fair subject for indictment as a dangerous nuisance. Dr. Simon says: " I accordingly think it an essential principle that the evil of a stinking sewer should always be dealt with at its root. Thus, a sewer which is imperfectly ventilated should have perfect ventilation provided for it; a sewer which, though fairly constructed, is from poorness of current not completely self-scouring, should at due intervals have extrinsic flushing; and sewers which, with radical ill-construction, are virtually but cess-pools under the street, should, without delay, be abolished."

Frequently, when the systematic sewerage of a town is undertaken, there comes up the question of private drains, which have been built by individual enterprise and are really the property of private owners; but owing to this complication, and to the fact that they are thought to be good enough for temporary purposes, they are often left to the last.

This is entirely wrong. *So far as circumstances*

will permit, the first action of the authorities should be to stop all connection of house drains with these sewers. The next should be to stop all connection of house drains with private cess-pools. This may seem, to those who have not considered the subject, like an extreme statement; but all who have studied the evidence as to the means of propagation of infectious diseases will recognize its justice. The health of the community would really be less endangered if the offensive matters sought to be got rid of were allowed to flow, in the full light of day, and with free exposure to the diluting air, in roadside gutters, than it now is by their introduction into the soil from which the water of house wells proceeds, and by the accumulation of putrefying masses in unventilated and leaky caverns, whence the poisonous gases sure to be produced find their way through the drains into our houses, or into their immediate vicinity. In the open air, their offensiveness would make us avoid them, and their poisonous emanations would be dissipated in the atmosphere. In the cess-pool and in a leaky sewer (which is but an elongated cess-pool) they too often find only one means of escape — through the drains into houses.

It is an almost invariable rule, in this country, to hold the question of sewerage in abeyance until some time after a public water supply has been provided. This is in every way unwise. It is a more than sufficient tax upon the soil of any ordinary village to receive its household wastes and subject them to a slow process of oxidation, so as to keep them, even under the most favorable circumstances, from doing great harm; but when the volume of these wastes is enormously increased by the liberal use of water from public works running free in every house, the case becomes at once serious. The soil is oversaturated, not only with water, but with water containing the most threatening elements of danger.

On the other hand, no system of sewerage *arranged to accommodate an abundant water supply* should be introduced until enough water is provided to secure the thorough cleansing of the drains.

Both branches of the work should be carried out at once, so that the oversaturation of the ground and the danger of sedimentary deposits in the sewer may alike be avoided. Even where the introduction of water is not contemplated, the local authorities of towns and villages should regard it as their most important duty to provide and maintain sufficient and absolutely impervious sewers wherever these are needed.

Nor is the simple foul-water drainage enough, save where the soil is so dry as to be free from such sources of malaria as do not depend on the wastes of human life. Malaria is a poison in the atmosphere which is recognized only by its effects on health. It often accompanies foul-smelling gases, but it is not necessarily heralded by any form of appeal to the senses, unless it be in the way of nervous headaches and a general feeling of debility.

Its presence is often marked by a disturbance of

sleep, uneasiness, lassitude, and digestive irregularity. Sir Thomas Watson, who has made one of the best statements of the case, says : "For producing malaria it appears to be requisite that there should be a surface capable of absorbing moisture, and that this surface should be flooded or soaked with water and then dried; the higher the temperature and the quicker the drying process, the more plentiful and the more virulent the poison that is evolved."

If malaria come from cryptogams, then drainage may prevent the germination of these, just as it prevents the germination of the seeds of the cat-tail flag.

The districts soaked by hill-waters about Rome were malarious for many centuries. Tarquin, by a system of deep subterranean drainage, collected this stagnant water and turned it into the Tiber. The lands became at once healthy, and were occupied by a large population. After the Gothic invasion the drains were neglected, and became obstructed, and so they still continue; and for hundreds of years these once fertile and populous districts have remained almost uninhabitable.

In addition to the frequent examples of sanitary drainage in Europe, and conspicuously in England, there are some instances in our own country which are sufficiently striking.

The town of Batavia, in New York, became at one time so malarious that it was almost threatened with destruction. It was decided to drain some saturated lands near the town. The first work was carried on by subscription, but the agricultural profit demonstrated was enough to induce land-owners to continue it at their own expense. The malaria was immediately mitigated, and for the past twenty years the town has been practically free from it.

Shawneetown, in Illinois, was formerly exceedingly unhealthy. One seventh of the men engaged in building the railroad there died of malarious disease. The draining of the surface water by a ditch (which at one point had to be cut to a depth of forty feet) removed the cause of the difficulty, and the town has remained healthy ever since.

Embryo towns and paper cities — their surface being obstructed by partly finished roads, and the land being withdrawn from cultivation and left to the care of no one in particular — are often much more unhealthy than their sites would have been had the same population planted itself in the open fields.

Stagnant pools on which cryptogams grow are frequent sources of disease. Most surface ponds have their areas contracted in summer by evaporation, and their newly-exposed, foul margins are quite sure to poison the atmosphere.

The increase of population in malarious districts always exerts an especially bad influence, because the organic wastes of human life accumulate in the soil and aggravate its insalubrity.

Closely allied to the malarious influences of saturated soils (especially in densely built districts) are those which attend the escape of sewer gas. The pernicious action of this gas is especially felt in the higher districts of sewered towns. As a rule, sewer air finds its escape in the higher-lying districts, and often conveys the germs of diseases originating in the lower and poorer parts of the town.

The medical officer of Glasgow says: "It has been conclusively shown that houses presumed to be beyond suspicion of any possible danger from this cause — houses in which the most skillful engineers and architects have, as they believed, exhausted the resources of modern science — have been exposed in a high degree to the diseases arising from air in contact with the products of decomposition in the sewers. And this for a very obvious reason. Such houses are usually built on high levels, where the drains have a very rapid fall."

Thon says that in Cassel, in the higher part of the town, which one would suppose the healthiest, typhoid fever was brought into houses by sewer gas which rose to them by reason of its lightness. In Oxford, in 1850, cholera, by the same action, appeared in several houses in the higher and healthier parts of the town.

In Berlin, in 1866, in those parts of the city where there were no sewers or water-closets, the deaths amounted to 0.37 per cent. of the population, while in the Louisenstadt, where sewers and water-closets were in general use, the deaths reached 4.85 per cent. Owing to errors in the con-

112 SANITARY DRAINAGE OF HOUSES AND TOWNS.

struction of the sewers of Croydon (England), their early use was followed by a violent outbreak of typhoid fever, which attacked no less than eleven per cent. of the population.

All experience and all scientific demonstrations go to show that the only safety in the water carriage system of sewering lies in the freest possible ventilation of the sewer and all its branches.

The evidence is almost universal, that wherever sewerage works are badly executed, and where proper precautions against the invasion of houses by sewer gas are not taken, typhoid fever and other diseases of the bowels are quite sure to be increased in intensity, and to appear in parts of the town which, before sewerage was undertaken, were comparatively healthy.

In 1856 there was an epidemic of typhoid fever in Windsor, England. Four hundred and forty persons, or five per cent. of the whole population, were attacked, and thirty-nine died. The disease affected the rich quite as much as the poor, but it confined itself entirely to houses that were in communication with a certain defective town drain. Windsor Castle had its own drain, and its inmates were entirely untouched; in the town, places only a block apart suffered severely or escaped entirely according as they were in communication with the town drain or with the castle drain.

It should be understood that sewage matters, though offensive, are not dangerous until two or three days after their production. The great point sought to be gained in the water system of sewerage, and that which constitutes its chief claim to confidence, is the instant removal of all organic refuse, everything being carried entirely away from the vicinity of the town before decomposition can have begun. Any plan not effecting this is entirely inadequate, and, on sanitary grounds, objectionable.

In many towns where there is no water supply, a rude system of sewerage is adopted, with the precaution of prohibiting water-closet connections. This is really hardly a precaution at all. Investigations made in towns where the earth and ash systems prevail, as in many of the large manufacturing towns of the north of England, show that the ordinary contents of the public sewers are in all respects not less foul and offensive, and probably little less dangerous, than are the contents of those which receive all of the ordure of the town with a copious flow of water. That is to say, the kitchen wastes and house slops when mixed with the wash of the streets constitute so prolific a source of offensive sewer gases that the night-soil is not especially marked, save as a specific vehicle for the spreading of such epidemics as are communicated by means of bowel-discharges.

On the other hand, Dr. Voelcker, who is excellent authority, does not accept the theory that sewage is as foul where house drainage is excluded from the sewers as where it is admitted. He says, " I do not think the Thames Conservators would have any objection to the surface water passing into the Thames through a separate sewer from the sewage. The statement that the proportion of the pollution of surface water to sewage is as ten to twelve must be founded on some mistake. I do not believe that water passing down the streets and running from the roofs of houses would naturally contaminate the river."

It is not the least benefit of the water supply in towns and villages that it sooner or later compels proper attention to the sewerage question; for a liberal supply of water, running free of cost in every house, soon leads to a great increase in the amount of water used and allowed to run to waste, and the result is that the people are awakened to the only argument by which average communities are at all affected, — the argument of life and death, — and are compelled, often in spite of themselves, to adopt more complete sewerage. It would show a wiser forethought, and lead to ultimate economy, if our towns would at once, on agitating the question of the introduction of water, couple with the scheme a plan of complete sewerage. It is a very ostrich-like blindness which hopes to escape the sure consequence of the beginning of the work. If it is undertaken at all, the double expense is inevitable, and it had better be honestly acknowledged and sufficiently provided for at the outset, especially as it is in every way better that the two operations should proceed simultaneously.

The question of cost should be taken into very early consideration, and it will not be slight; but *pari passu* there should be a due estimate of the benefits to accrue. These are not of such a character that they can be very readily calculated in dollars and cents, but there few cases, in towns of five thousand inhabitants and over, where their importance will not be very fully appreciated.

The construction of a proper system of sewerage is at best expensive, but it may be much more cheaply done if taken in hand at once and carried on systematically until the whole is complete, than if done piecemeal, here and there, as property-holders may elect, which is the general custom in America. I do not know that the English method of paying for the cost by distributing principal and interest over a period of years has been adopted with us, but it seems the most just and the least oppressive. It is more fair to posterity, without bearing heavily on the present generation, than payment by interest-bearing bonds to be redeemed twenty or thirty years hence.

Latham, in his inaugural address as President of the Society of Engineers, made a calculation of the cost and value of the water-works and sewerage of the town of Croydon, as follows: —

Cost: purchase of land (for sewage utilization), £50,000; water-works, £70,000; sewers, irrigation works, baths, abattoirs, and general improvements, £75,000. Total, £195,000. The money savings during thirteen years since the completion of the work, he estimates to have been: 2,439 funerals, which would have cost £12,195; 60,975 cases of sickness prevented, £60,975; value of the labor for six and one half years of 1,317 adult persons whose lives were extended, £166,930. Total, £240,100. He says, "Although it has been attempted to put a money value on human life, we individually feel that life is priceless, and we may look to the 2,439 persons saved from the jaws of death in this single town as the living testimony of the great value of sanitary works."

It is well known to physicians that their chances of success in the treatment of disease are very much reduced with persons living in unhealthy places.

The cost of sewerage works is often made unnecessarily great with the idea that it is the duty of the public to furnish on outlet for factories, slaughter-houses, and all manner of establishments which are carried on for individual profit, and in which the cost of removing the resultant refuse is fairly chargeable on the business rather than on the public purse.

So far as the community is concerned, it should be compelled to construct sewers only for the removal of such waste matters as are incident to the daily life of all classes of the population. If breweries, chemical works, and other manufactories producing a large amount of liquid waste, are to be provided with a means of outlet, this should be done entirely at their own charge; their profit and convenience should not be advanced at the cost of every member of the community. And more than this, the wastes of factories being often pernicious, not only on reaching the outlet of the sewer, but by the generation of gases within them which may pervade all their ramifications, it is a serious question whether such establishments should not be compelled to secure independent outlets at their own expense, or at least to render their wastes innoxious before discharging them into the public drain; paying even then an extra sewer-rate, proportionate to the extra service they require.

The sanitary authority of every town should have entire control over the sewers, with power to decide what shall be admitted to them, and what excluded, and to levy an additional tax in all cases where an undue use is made of the public convenience.

The economical use of the organic wastes of the house or town, demands most careful consideration. The utilitarian question, important though it is, is only secondary, but as an accessory, the matter of economy is very important, and in every *perfect* system of sanitary improvement the arrangements must be such that there shall be a complete utilization of all the valuable constituents of the wastes of domestic life; and practically our arrangements should be so nearly perfect, that nothing shall be lost that can be economically saved.

The more important considerations affecting the question of town sewerage, were stated in the

118 SANITARY DRAINAGE OF HOUSES AND TOWNS.

"general conclusions" of the English Board of Health, after a thorough investigation of the whole subject of sewerage, as follows: —

1. That no population living amidst aerial impurities arising from putrid emanations from cesspools, drains, or sewers of deposit, can be healthy or free from attacks of devastating epidemics.

2. That as a primary condition to salubrity, no ordure or refuse can be permitted to remain beneath or near habitations, and by no other means can remedial operations be so conveniently, economically, inoffensively, and quickly effected, as by the removal of all such refuse dissolved or suspended in water.

3. That the general use of large brick sewers has resulted from ignorance or neglect; such sewers being wasteful in construction and repair, and costly through inefficient efforts to keep them free from deposits.

4. That brick and stone house drains are "false in principle, and wasteful in the cleansing, construction, and repair. . . . . That house drains and sewers, properly constructed of vitrified pipe, detain and accumulate no deposit, emit no offensive smells, and require no additional supplies of water to keep them clear."

5. That an artificial fall may be cheaply and economically obtained by steam pumping, and that the cost of the whole system to each house is much less than the cost to that house of removing its refuse by hand.

THE DRAINAGE OF TOWNS. 119

6. All offensive smells proceeding from any works intended for house or town drainage, indicate the fact of the detention and decomposition of ordure, and afford decisive evidence of malconstruction, or of ignorant or defective arrangement.

PRACTICAL DIRECTIONS.

CHAPTER IV.

ARRANGING PLANS FOR TOWN SEWERAGE.

WHETHER it is contemplated to execute the whole work of sewering of towns immediately or not, the first step in every case, should be to prepare a complete plan of the whole work, so that when it shall finally be finished it will be harmonious, — each part being adapted to the work that it will have to perform, when all the lines are in operation. In arranging this plan, the engineer will consider, not only what are to be the demands of the town as it now exists, but in what way the sewage of parts to be built in the future will be likely to affect the demand upon its main lines. Of course it will be impossible to foresee with precision the extent and direction of the future growth of any town, and this element of uncertainty must always remain. Still, so far as the probabilities of the case are concerned, much economy can be secured by a careful consideration of the prospect, — providing for rather more than less of what will probably be necessary, but arranging so far as may be that these parts to be added shall not all demand an outlet through the same main line. In this way we may avoid the necessity for making any sewer very much larger than present needs require,

124 SANITARY DRAINAGE OF HOUSES AND TOWNS.

while by increasing the size of several mains we may provide all of the adjacent area when the occasion shall arise. So far as future extensions of the present town are concerned, all that need be done is to provide main outlets, of a size adapted to their area.

OUTLETS. — INTERCEPTING SEWERS.

The first question to be considered in arranging the plan for the sewerage of a town or village is that of an outlet, at which the foul sewage of the streets and houses may be delivered without danger of polluting water-courses or destroying their fish, or of silting up harbors or navigable streams; and without forming within dangerous proximity to the town a deposit of offensive sewage matters which might constitute a source of annoyance or of insalubrity.

In all cases where this part of the problem presents difficulties, it should be considered whether a separate direction or a shorter outfall may not be given to the storm-water drainage, allowing the sewers to deliver at their main outlet only the ordinary drainage of houses and the street-wash of very slight rains. The cases are frequent where the removal of the sewage proper from low lying parts of a town may be best and most economically secured by artificial pumping; though, in the majority of instances, it will be practicable, by the use of intercepting sewers, to deliver by natural outfall the drainage of all except the very lowest portions of the town. It is in the adjustment of this part of the work that the experience and judgment of the engineer in charge will be the most severely tested; in all matters of construction, ventilation, house connections, etc., certain rules and explicit directions can be applied, but the arrangement of the outlet varies with nearly every new undertaking, and with reference to this branch of the subject it is possible here to give only general indications.

It would often be practicable to take the small ordinary flow of public sewage to a remote point, by the use of an intercepting sewer, even when the cost of providing such an outlet for *storm water* would be so great as to make it impracticable. In such cases there may be carried from the point of outlet to the distant point of discharge the smallest pipe that will accommodate the usual flow, so arranged that whenever, as during storms, the volume is increased beyond the capacity of this pipe, it shall overflow and be carried directly into the stream or harbor at hand. At such times the amount of water in the sewage will so dilute it that no bad effect need be apprehended.

The great danger in nearly all the towns of our Atlantic seaboard lies in the fact that they discharge some of their most important sewers below high-water mark, so that at each rise of tide not only is the flow at these points checked, and foul silt allowed to collect in the stilled water, but the closing of the vent at this end of the sewer and the rise of water within it, whether by the action of the tide or through the accumulation of the flow from above, brings a pressure to bear upon the contained air and forces it to escape at the higher points; the state of the tide is in this way often made perceptible by the forcing of water traps a mile or more distant from the outlet.

Outlets, especially of large sewers, exposed to strong winds, are likewise very objectionable, the pressure of the wind forcing the tainted air to find vent too often through badly trapped drains leading into occupied houses.

Where necessary to secure a constant flow of sewage, pumping should always be resorted to, to avoid the expedient, now often adopted, of using some part of the system for the temporary storage of sewage during high tides. With coal at nine dollars per ton, the cost of lifting thirty thousand gallons ten feet high with a twenty-five horse-power engine would not exceed seventy-five cents, while with a larger engine and a larger flow the relative cost would be much less. It was estimated that to lift the whole sewage and rain-fall from a low-lying district in London, occupying four thousand acres, to a height of thirty-one feet would cost about five cents per annum per head of population. Whatever the cost of pumping, it may be made in level districts to do away with any outlay for cleansing or flushing sewers, which without pumping must have been laid nearly level.

There are as yet few cases in this country where it is necessary to discharge the sewage of a town into a stream from which other towns receive their water supply, though the towns along the Schuylkill River still stand in this relation to the city of Philadelphia. The time is probably not very distant when this question will become here, as it now is in England, a very serious one.

Tidal estuaries and bays receiving the drainage of a town are sure to have those parts of their bottoms and sides which are alternately covered and exposed by the changing tides fouled with organic matter, and to become thereby seriously offensive and dangerous.

Recent sewage floats in water. After maceration it sinks in still water and in currents having a less velocity than one hundred and seventy feet per minute. Its specific gravity is about 1.325.

The condition of Newtown Creek, Wallabout Bay, and the Gowanus Canal and Bay, near Brooklyn, are examples of the subsidence of sewage in eddies and slack water.

Tides may be made extremely useful in the flushing of sewers in level lowlands, but care should be taken to carry the outlet to a point where the inconvenience from subsidence will be reduced to the minimum.

SIZES OF SEWERS.

Nearly the most important item in connection with the arrangement of a plan for sewerage, and one in which professional experience is especially important, is the regulation of the sizes of the different main drains and laterals. This involves a consideration of the amount of sewage proper; the customary rain-fall of the district; the grade or inclination of the surface, as indicating the rapidity with which storm waters will find their way to the entrances of the sewers; and the extent to which, in order to avoid the flooding of cellars and other injury during copious rains, it is advisable to increase the sizes of the conduits beyond what is needed for ordinary use.

It is doubtful whether even large cities can really afford, in arranging their sewerage, to provide for the underground removal of the water of heavy rains, and certainly in smaller towns and villages it would be far cheaper to pay for repairing whatever damage might be caused by occasional heavy floods in the streets, or to provide for the removal of the water of these storms by surface gutters, than to make the size of the whole system of sewerage adequate for such work. Not only this, but sewers large enough to accommodate the water of very heavy storms would usually be too large for perfect cleansing with their daily flow, and would require expensive flushing appliances, which with smaller pipes would not be needed. In country towns it would not generally be wise to provide for removing through the pipes the flow of a heavier storm than one quarter inch per hour. Gutters are much cheaper than sewers, and there is usually no objection to their being relied on to remove the surplus water of sudden showers.

It is not unusual to provide in cities for a rainfall of one inch per hour, and to assume that one half of this will reach the sewer within the hour. Even this is far more than is necessary, if any other provision can be made for exceptional storms. For example: In Providence, one hundred and eighty-five storms were recorded in twenty-six years. Of these only twenty-seven were of more than one half inch, and one hundred and thirty-one of them were of one fourth inch or less. One half inch per hour equals thirty and one fourth cubic feet per minute per acre.

If the supply of water in a town is ten gallons per head per day, for the whole population, the quantity of sewage to be removed will be about one hundred pounds daily for each person. Of this the closet flow will constitute about one third. This assumes that the use of the water-closet is universal, that vaults are entirely done away with, and that the water is employed for all domestic requirements.

In Brooklyn, it is estimated that, aside from rain the sewage equals one and one fourth times the water supply, or fifty million gallons per day, the half of which running off between nine A. M. and five P. M. gives 3,125,000 gallons per hour, escaping during eight hours. This, from twelve hundred acres, gives two hundred and sixty gallons or thirty-three cubic feet per acre per hour, being less than one hundredth of an inch in depth over the whole area.

Mr. Shedd, engineer to the city of Providence,

says on the subject of the size to be given to a sewer: " The capacity of sewers to carry water depends mainly on their sectional area — if of proper form — and the rate of fall. In order to render sewers as nearly self-cleansing as possible, they must, as before stated, be adapted, in size and inclination, to the ordinary flow of sewage, so far as to keep up a velocity sufficient to carry on all light matters, and to leave only so much heavy matter as will be finally carried along by the scouring effect of the storm waters. . . . .

" There is room for question as to how far the sewer should be made capable of carrying extraordinary storms. The original cost of large sewers, as well as the cost of maintenance, is so much greater, that a city can afford to pay for damages done by storms of unfrequent occurrence, rather than to construct them. Just where to stop in providing for such storms is a matter of doubt; and what would secure true economy in one place would not necessarily do so in another. The frequency of the heavy storms, the amount of rainfall, and the damage likely to be done, depend upon the location and the circumstances of each place."

Very careful records of rainfalls were kept for a long time by President Caswell of Providence, showing the following tabulated result which gives in an interesting form the data that are of value in sewerage work.

" In twenty-six years previous to 1860, the time of rainfall is recorded in 185 storms : —

ARRANGING PLANS FOR TOWN SEWERAGE.

In 131 storms, rain fell at the rate of 0.25 inch per hour, or less.
In 18 storms, rain fell at the rate of 0.33 inch per hour, about.
In 9 storms, rain fell at the rate of 0.40 inch per hour, about.
In 7 storms, rain fell at the rate of 0.50 inch per hour, about.
In 8 storms, rain fell at the rate of 0.62 inch per hour, about.
In 3 storms, rain fell at the rate of 0.67 inch per hour, about.
In 3 storms, rain fell at the rate of 0.75 inch per hour, about.
In 4 storms, rain fell at the rate of 0.87 inch per hour, about.
In 1 storm, rain fell at the rate of 1.00 inch per hour, about.
In 1 storm, rain fell at the rate of 1.75 inch per hour, about.
185

"The two storms giving 1 inch and 1¾ inch per hour fell on the 5th and on the 14th of July, 1841.

"In one case, where the time of rain-fall was less than an hour, the rate is made as though the rain was an hour in falling.

"In fourteen years, to the 1st of January, 1874, the time of rain-fall is recorded in 139 storms:—

In 98 storms, rain fell at the rate of 0.25 inch per hour, or less.
In 9 storms, rain fell at the rate of 0.33 inch per hour, about.
In 2 storms, rain fell at the rate of 0.40 inch per hour, about.
In 10 storms, rain fell at the rate of 0.50 inch per hour, about.
In 5 storms, rain fell at the rate of 0.60 inch per hour, about.
In 3 storms, rain fell at the rate of 0.70 inch per hour, about.
In 2 storms, rain fell at the rate of 0.80 inch per hour, about.
In 1 storm, rain fell at the rate of 0.90 inch per hour, about.
In 1 storm, rain fell at the rate of 1.00 inch per hour, about.
In 1 storm, rain fell at the rate of 1.12 inch per hour, about.
In 1 storm, rain fell at the rate of 1.20 inch per hour, about.
In 1 storm, rain fell at the rate of 1.40 inch per hour, about.
In 1 storm, rain fell at the rate of 1.52 inch per hour, about.
In 1 storm, rain fell at the rate of 1.83 inch per hour, about.
In 1 storm, rain fell at the rate of 2.00 inch per hour, about.
In 1 storm, rain fell at the rate of 2.32 inch per hour, about.
In 1 storm, rain fell at the rate of 3.15 inch per hour, about.
139

"Of those storms giving an inch or more per hour,—

One fell in 1862.
One fell in 1863.
One fell in 1868.
One fell in 1869.
Three fell in 1870.
One fell in 1871.
One fell in 1873.

Making nine such storms in fourteen years, against two in the previous period of twenty-six years. Seven of these storms occurred in the last six years.

" Where the time of rain-fall was less than an hour, it is reckoned as having been an hour in falling."

During my direction of the draining work of the Central Park in New York, there was gauged (on the 13th of July, 1859) a sudden shower in which between 5.15 P. M. and 5.45 P. M. two inches of rain fell.

On the 30th and 31st of October, 1866, the Croton Aqueduct Department gauged a rain storm in New York which lasted for about five hours and measured four inches of rain.

Mr. Rowe, for a long time superintendent of sewers in the Holborn and Finsbury District of London, writing in defense of large sewers which he had constructed, says : —

" I have observed twenty-five cubic feet of water per minute per acre reach the sewers from an inch fall of rain in the hour, from a surface where the houses have much garden ground attached; and in another case, where the houses were nearer together, thirty-three cubic feet per acre per minute. That greater falls of water do take place, and that not unfrequently, is a well-known fact. I have known ten instances of the kind during the period of my observations " (twenty years).

ARRANGING PLANS FOR TOWN SEWERAGE.

It is a safe rule to estimate all sewage except rain-fall at eight cubic feet per head of population per day. Of this, one half will be discharged between nine A. M. and five P. M., equal to a flow of five hundred cubic feet per hour for each thousand of the population.

Sewers choke and overflow during heavy storms mainly because they are too large for the work they are ordinarily called on to perform. If a sewer is so small that its usual flow is concentrated to a sufficient depth to carry before it any ordinary obstruction, it will keep itself clean. But if, as is almost always the case where the engineer lacks experience or where he defers to the ignorance of the local authorities, it is so large that its ordinary flow is hardly more than a film, with no power even to remove sand, we may be quite sure that its refuse solid matters will gradually accumulate until they leave, near the crown of the arch, only the space needed for the smallest constant stream. And, in order to make room for a rain-fall flow, the whole

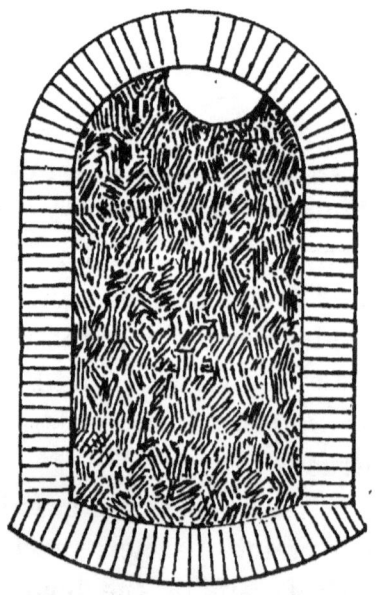

Figure 2. — Cross section of a large sewer filled by the gradual accumulation of silt until only sufficient water-way is left for the smallest constant flow.

sewer will have to be cleared by the costly and offensive process of removal by manual labor. A smaller sewer would have been kept clear by its own flow.

The shallower and broader the stream, the more the friction against the bottom and sides and the greater the retarding of velocity. A brick will stand unmoved in a shallow stream of water running sluggishly through a fifteen-inch drain, while if the same stream were concentrated into a five-inch drain it would have so much greater depth, force, and velocity, that the brick would be entirely covered and swept away.

The passion for too large pipes seems to be an almost universal one. The feeling is that it is best to make the conduit " big enough anyhow," and as a result, nearly every drain that is laid, in town or country, is so much larger than is needful that the cost of keeping it clean is often the most serious item of cost connected with it.

One principle is very apt to be disregarded in regulating the sizes of sewers; that is, that after water has once fairly entered a smooth conduit having a fall or inclination towards its outlet, the rapidity of the flow is constantly accelerated up to a certain point, and the faster the stream runs the smaller it becomes; consequently, although the sewer may be quite full at its upper end, the increasing velocity soon reduces the size of the stream, and gives room for more water. It is found possible, in practice, to make constant additions to the volume of water flowing through a sewer by means of inlets entering at short intervals, and the aggregate area of the inlets is thus increased to very many times the area of the sewer itself. Where a proper inclination can be obtained, a pipe eighteen inches in diameter makes an ample sewer for a population of ten thousand.

It was formerly the custom with architects and engineers to enlarge the area of any main pipe or sewer in proportion to the sectional area of each subsidiary drain delivering into it. But this is no longer done, since it has become known that additions to the stream increase its velocity, so that there is no proportionate increase of its sectional area. For example, the addition of eight junctions, each three inches in diameter to a main line of four-inch pipe, did not increase the sectional area of its flow, but made the flow only more rapid and cleansing. Ranger thus illustrates the average architect's

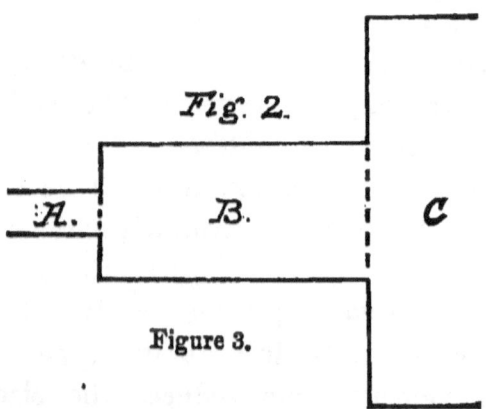

Figure 3.

A, 3-inch drop or soil pipe.
B, 9-inch intermediate drain (9 times the area of *A*).
C, 26-inch sewer (8½ times the area of *B*, and 75 times the area of *A*).

method of draining a house and court. The reason for making *B* so large is to *prevent* its choking, an effect that its extra size is quite sure to produce.

The main sewer in Upper George Street, in London, is five and one half feet high and three and one half feet wide. In the bottom of this sewer there was laid a twelve-inch pipe five hundred and sixty feet long. A head-wall or dam was built at the upper end of this, so that all the sewage had to pass through the pipe. The whole area drained was about forty-four acres (built area). The velocity of the water in the pipe was found to be four and one half times greater than on the bed of the old sewer. The pipe contained no deposit, and during rains stones could be heard rattling through it. The force of water issuing from the pipe kept the bottom of the old sewer perfectly clean for about twelve feet below its mouth. From this point bricks and stones began to be deposited, and farther on sand, mud, and other refuse, to the depth of several inches. In one trial a quantity of sand, bricks, stones, mud, etc., was put into the head of the pipe; the whole of this was passed clear through the pipe, and much of it was deposited on the bottom of the old sewer some distance from its end. *The pipe was rarely observed to be more than half full at its head.* It was found that the sum of the cross sections of the house drains delivering to this half-full twelve-inch pipe was equal to a circle thirty feet in diameter.

Another experiment was made with a sewer in

Earl Street, which took the drainage from twelve hundred average-sized London houses, the area occupied being forty-three acres of paved or covered surface. It was three feet wide and had a sectional area of fifteen feet, with an average fall of one in one hundred and eighteen. The solid deposit from the twelve hundred houses accumulated to the amount of six thousand cubic feet per month (two hundred and twenty-two cart-loads). A fifteen-inch pipe placed in this sewer, with an inclination of one in one hundred and fifty-three, kept perfectly clear of deposit. The average flow from each house was about fifty-one gallons per day, and, apart from rain-fall, the twelve hundred houses would have been drained by a five-inch pipe. It was estimated that at that time (about twenty-five years ago) the mere house drainage of the whole of London might be discharged through a sewer three feet in diameter; yet there is probably not a village of five thousand inhabitants in the United States whose magnates would be satisfied with a sewer of much less size for their own purposes; and a single hotel in Saratoga has secured future trouble in the way of the accumulation of raw material for the production of poisonous sewer gas, by laying a drain for its own use thirty inches in diameter.

A fifteen-inch sewer was formerly considered the smallest size admissible for the drainage of a "mansion." Such a sewer, with a fall of one in one hundred and twenty, or one inch in ten feet, would drain nearly two hundred of the largest city houses;

and a nine-inch drain with the same inclination would remove the house-drainage *and storm water* from twenty such houses.

A curious example of the capacity of small pipes was furnished in a case where a six-inch pipe was laid for the drainage of one detached house. One after another, as new houses were built, new drains were connected with this same pipe, until, after a time, it was found to be clean and in perfect action, though carrying all the drainage of one hundred and fifty houses. In a second instance a workman by mistake used for the drainage of a large block of houses a pipe which the architect had intended for a single house, and it was found to work perfectly.

It may be taken as a rule that, with even a slight fall, a well-constructed eighteen-inch pipe sewer is ample for the drainage of an ordinary village area containing seven or eight hundred houses. In one instance a sewer of this size, having a fall of one in one thousand, accumulated but little deposit, and this was always removed by storms. In Tottenham (London), a main sewer of nine-inch pipe, widening to twelve-inch and afterward to eighteen-inch, and having a fall of one in one thousand and sixty-two, drained an area containing sixteen hundred houses. Its ordinary current was two and one half miles per hour, and brickbats introduced into it were carried to the outlet. During ordinary continued rains it was not more than half full half a mile from the outlet, and at the outlet the stream was only two and three fourths inches deep.

Rats and vermin live and breed in large sewers, never in small pipes.

While these chapters were being prepared for publication, the Sewer Commissioners of Saratoga (the writer being employed as their consulting engineer) completed a main sewer more than two miles long, for the removal of the entire sewage, rain-fall, and spring-water drainage of that village. The experience with this work affords so pertinent an illustration of the principles here advanced that it seems worth while to refer to it. The village is large and scattered, has an abundant water supply, is so inclined that during showers its storm waters concentrate rapidly, and has, aside from its regular population, five or six enormous hotels, entertaining, when full, about as many thousand guests. The village brook itself, being mainly supplied by spring water flowing from various points over a wide district, is always a considerable stream. As it flowed through its old channel — a conduit with rough, loosely-laid stone side-walls, and with a more or less irregular bottom — its sectional area was about five feet. During heavy rains it was sometimes thrice this.

From the very beginning of the work we encountered the most violent opposition on the part of many citizens, who believed that the sewer contemplated (circular, three feet in diameter) would be entirely inadequate, not only for the removal of the water of heavy rains, but even for the drainage of the hotels alone, or the carrying of the storm waters alone; and throughout its construction this main sewer was derided as a "cat hole." We were constantly reminded that one hotel had a main drain eighteen inches in diameter, and another a drain two and one half feet in diameter, and that it was madness, with these drains as our guide, to attempt to accomplish the whole work with a three-foot sewer; especially as our fall was said to be slight, one foot in four hundred feet.

On the 9th of July, 1875, the connections were made with all of the hotels; the village brook itself was turned into the sewer at its head, and its insufficiency was to be demonstrated. After every available source of water had been drained, the depth of flow in the upper part of the sewer was six and one half inches. Nearer the outlet, where the water had acquired its maximum velocity, it was only four and one half inches. As this was not sufficient to wash out the few loose boards carelessly left by the workmen who had done the final pointing of the joints, a hydrant was turned on at the upper end of the sewer, with a full head, and it had the effect only of raising the flow one inch at the upper end and less than half an inch at the lower end of the sewer. On the 10th there fell a violent thunder-shower, flooding the street gutters until the water ran to the top of the curb-stones, and when this flood had reached the catch-basins and the open brook that discharged into the head of the sewer, its only effect was to raise the flow, at the highest point, less than two inches, justifying the original opinion that a

ARRANGING PLANS FOR TOWN SEWERAGE. 141

two foot sewer would have been more than adequate for all that was required of it. On the 30th day of August the entire village brook, with its tributaries and its many springs, was turned into the three-foot sewer, near the water-works, about one-half mile beyond the outskirts of the village. The effect of this addition was to increase the depth of flow in the sewer from about six inches to nine inches, and to increase the velocity of its stream from one hundred and fifty feet per minute to one hundred and eighty-five feet per minute. I can excuse my course in recommending so large a sewer as one of three feet, only by the fact that in the state of public opinion then it would have been entirely impossible to secure the making of anything smaller. Before the introduction of the brook I examined the outlet of the Grand Union Hotel, which had then about eight hundred and fifty guests and four hundred and fifty servants, or about thirteen hundred inmates in all. There can hardly be fewer than one hundred water-closets in the house, and the use of water in this hotel seems to be in every way as copious as possible. The hour of examination was ten in the morning, at which time, as the landlord supposes, the largest flow is running. By the most careful measurement and estimate that I could make, the amount of sewage then flowing from that hotel measured four and one half inches in sectional area, and might have all been discharged by a two and one half inch pipe.

Concerning the rate of fall necessary for the removal of ordinary road silt from sewers, Adams gives the following table of inclination for pipes of different sizes *running half full;* based on careful calculations and practical trials with the sewerage works of the city of Brooklyn.

>For 6-inch pipes a grade of 1 in 60.
>For 9-inch pipes a grade of 1 in 90.
>For 12-inch pipes a grade of 1 in 200.
>For 15-inch pipes a grade of 1 in 250.
>For 18-inch pipes a grade of 1 in 300.
>For 24-inch pipes a grade of 1 in 400.
>For 30-inch pipes a grade of 1 in 500.
>For 36-inch pipes a grade of 1 in 600.
>For 42-inch pipes a grade of 1 in 700.
>For 48-inch pipes a grade of 1 in 800.

When the direction changes, the friction is increased, and the fall must be increased to compensate for this.

When the lay of the land permits it, the most rapid fall should be given at the upper end of the sewer, where the quantity of water is least, and where the greatest velocity is consequently needed to secure a cleansing flow.

The object of giving an inclination or fall to the sewer is to secure the velocity necessary for the removal of such solid matters as may exist in the sewage, but *if the amount of water flowing is proportionate to the size of the conduit*, sewers of different sizes give the same velocity at different inclinations: for instance, a ten-foot sewer with a fall of two feet per mile, a five-foot sewer with a fall of four feet per mile, a two-foot sewer with a fall of ten feet per mile, and a one-foot sewer with a fall of twenty feet per mile, will have the same velocity, provided they are filled in proportion to their capacity; but the ten-foot sewer will require one hundred times as much sewage as will the one-foot sewer, *and unless it carries a volume of water proportioned to its capacity, the velocity of its stream will be correspondingly lessened.* It becomes, therefore, especially important that the *size of the conduit* be adjusted to the *volume of the stream*, this being as important as the rate of inclination in securing a cleansing flow, and being so little understood that it cannot be too much emphasized in any attempt to bring the mechanism of sewerage works to the notice of the general public.

Latham gives a velocity of three feet per second as the least that should be allowed for the outlet drain of a house. A four-inch drain to secure this flow should have a minimum inclination of one in ninety-two; a six-inch drain, one in one hundred and thirty-seven; a nine-inch drain, one in two hundred and six; and to attain a velocity of three feet per second at these inclinations *they must run at least half full;* that is, the four-inch drain must discharge 7.85 cubic feet per minute; six-inch 17.66 cubic feet per minute; and nine-inch, 39.76 cubic feet per minute. It is very seldom indeed that even a large boarding-house discharges a flow equal to 7.85 cubic feet per minute, and in practice, while too large outlets should aways be avoided for house drains, any such drain should have considerably more than the minimum rate of fall indicated above.

It cannot be too often reiterated that the great purpose of modern water sewerage is to remove immediately, entirely beyond the occupied portions of a town, all manner of domestic waste and filth before it has time to enter into decomposition; thus preventing an accumulation of dangerous matter, and obviating the necessity for employing men in the unwholesome work of hand-cleansing of cess-pools and of sewers of deposit, *which all sewers are apt to become when materially too large for the work they have to perform.*

The pipe sewer has been so long in successful use that there is no further question of its value. Even ten years ago, fifty miles of such pipe were made per week in Great Britain alone.

Accuracy in form and joints, and smoothness of surface, are very important. A perfectly round pipe, accurately laid at the joints, will deliver, under the same circumstances, fifty per cent. more water than one of distorted form or with ill-fitting joints.

Any roughness of surface as in even the best made cement pipes, tends to catch hair and lint and thus to form nuclei for accumulating obstructions, sometimes so hard that they can be removed only by forcible mechanical means.

With a well-constructed system of pipe sewers, not too large for the work required of them, of good form and surface, with perfect joints, with only curved junctions, and with a well regulated even if slight fall, every particle of the sewage of the town may be delivered at the outlet, far away from the built-up districts, long before any decomposition of the refuse matter has set in; though occasional flushing may be necessary to cleanse the sides of the pipes from slimy matters adhering to them.

The material of the pipe should be a hard, vitreous substance, not porous, since this would lead to the absorption of the impure contents of the drain, would have less actual strength to resist pressure, would be more affected by frost or by the formation of crystals in connection with certain chemical combinations, or would be more susceptible to the chemical action of the constituents of the sewage. The best pipe known in our market is the Scotch; but some American work is very nearly as good.

Much experience with cement sewer pipes seem to demonstrate that they are not sufficiently uniform in quality, nor sufficiently strong and durable to be used with confidence in any important work, — whether public or private.

Sewer pipes should be salt-glazed, as this requires them to be subjected to a much more intense heat than is needed for slip-glazing, and thus secures a harder material.

Pipes having a socket at one end should be furnished with a gasket before being cemented, in order that no cement may be pressed through into the bore of the sewer, to cause a disturbance of the flow. Where there is danger of the penetration of roots, as near elm-trees, the sewer should be bedded in a sufficient thickness of concrete to prevent the entrance of rootlets, which are sure to find and to penetrate the smallest aperture. An entrance once effected, a mass of fibres soon forms, sufficient to retard or entirely to arrest the flow.

There has been patented in England a joint for earthern-ware drains which is made by casting upon the spigot and in the socket of each pipe, by means of prepared molds, rings of good cement which will make a tight fit and bring the bore exactly into line.

A chief argument in favor of the use of pipes rather than brick sewers lies in their greater essential cleanliness. Brick sewers are always offensive, even though small, because their porous walls are more or less permeated by the filth of their contents. If (as is almost always the case) they are too large, there will be the additional annoyance of accumulations of refuse as foul and dangerous as the contents of any cess-pool, producing poisonous gases which are free to travel through the sewer and all its branches.

FORMS OF SEWERS.

The desirable forms of sewers are three in number: the round, the elliptic, or egg-shaped, and that with perpendicular walls, having a semi-circular roof and a hollowed floor.

The circular form gives the greatest sectional area for the amount of wall required, and, therefore, the greatest capacity for discharge. All sewers which are at all well adapted in their size to the regular daily work that they will have to perform are best made of circular shape, but where in addition to the daily use provision has to be made for the removal of the waters of heavy rains, so that at times a very much greater capacity will be needed, the elliptical form is to be preferred.

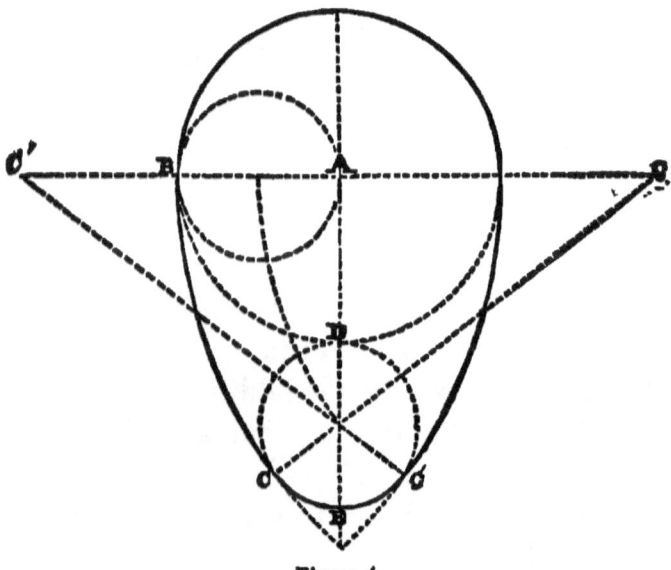

Figure 4.

Figure 4 illustrates the advantage of the elliptical form for this purpose. The circle $C\,D\,C'$ represents a circular sewer twelve inches in diameter, which we will suppose to be sufficient for the ordinary sewage of a district in which the minimum depth of flow would be three inches.

As there is ordinarily only this minimum flow, and as it is desirable to preserve this depth in order to cleanse the sewer, it would be unwise, in order to secure the storm water capacity desired, to adopt a large circular sewer in which, because of its greater width, the minimum flow would be less than one inch deep,— not sufficient in depth or velocity for the removal of deposits. Therefore, in order to increase the capacity as much as is necessary, instead of enlarging the circle we preserve the bottom form, $C E C'$, remove the top of the circle and carry it up to a height of thirty-six inches, its greatest width being twenty-four inches. In this way we preserve always our cleansing flow in a narrow and deep channel and give the needed capacity for the removal of storm waters.

Pipe sewers, which may be economically used up to a size of eighteen inches diameter, should always be round,— the slight warping to which earthenware is subjected in burning being likely to throw the ellipse out of its form, so that good joints cannot be made, while with the circle we can at least be sure of a perfect fitting in the water-way by turning the truest side of the pipe to the bottom. Larger sewers, if they have sufficient *regular* flow to carry off all sedimentary matters, should still be made round, as this form is the cheapest and the strongest.

It is only when the flow is very irregular, and when this is the only available means for securing the proper cleansing minimum flow, that the egg-shaped should be adopted.

ARRANGING PLANS FOR TOWN SEWERAGE. 149

The sewer with vertical sides and arched roof is to be adopted only under exceptional circumstances, when the minimum flow is always large, and where great capacity is needed.

THE SEPARATE SYSTEM.

In what has been said, the carrying away of the water of excessive floods to a separate point of outlet, giving a more remote or more artificial outflow to the regular sewage of the town, I am not to be understood as indorsing all that its advocates claim for what they call "The Separate System," which is (theoretically) the carrying of all rain-fall and all surface water away by one outlet, and the car-carrying of the foul sewage (house waste, etc.) through the regular system of pipes, delivering it in a concentrated form for agricultural use.

Whatever advantage may arise to the farmer from the fact that he receives his liquid manure in a more concentrated form, and that it comes to him in a regular daily quantity which he may more readily arrange to use, there would be a more than corresponding disadvantage to the public in the fact that the house waste alone is not sufficient, save perhaps where the grades are very steep, to keep the sewers clean. Under ordinary circumstances, an attempt to make this disposition of the sewage matters of a town would undoubtedly result in the necessity for much artificial flushing and cleansing, and to the danger of the frequent stoppage of smaller pipes and house drains.

150 SANITARY DRAINAGE OF HOUSES AND TOWNS.

To keep *all* rain-fall out of the public sewers is objectionable for more reasons than one, for no artificial flushing which can be depended on in practice can be so efficient in cleansing the sewers as the frequent introduction of a sufficient amount of rain water. On the other hand, where it is practicable to do so, the removal of the surface water of *excessive* storms by some channel entirely separate from the general system of sewerage has the great advantage of economy, while it often enables us to secure within reasonable cost the more distant removal of the foul sewage. (See note on page 171.)

THE VENTILATION OF SEWERS.

All sewers must at least be *vented*, and for perfect security all ought to be well ventilated. It is of the first importance to provide openings for the escape of the contained air and gases when these are compressed, either by a wind blowing into the outlet or by the increase of the quantity of water in the sewer from the rise of the tide or from heavy rain-fall. Unless such precaution is taken, house traps will surely be forced and sewer gas will surely escape into dwellings. It is, however, hardly less important that there should be such a free circulation of air through the sewer as will prevent the formation of those poisonous, mephitic gases which are especially generated in the absence of a sufficient supply of oxygen.

Latham says that unventilated sewers are far more dangerous than steam-engines without safety-

ARRANGING PLANS FOR TOWN SEWERAGE. 151

valves. They contain in their air some quality that is pestilential and dangerous to health, and this can be disposed of or neutralized only by giving the air of the sewer a free communication with the atmosphere. Typhoid fever is said rarely to be absent from towns with unventilated sewers. The constantly changing pressure upon the confined air within these conduits acts in connection with the draughts of chimneys and the force of winds to cause the bubbling of house traps, accompanied with an entrance of more or less of the sewer emanations.

When the sewerage works of Croydon were nearly completed and the town was visited by an epidemic of typhoid fever, the mortality rose from 18.53 per thousand to 28.57 per thousand. Although it is probable that the only matters decomposing in the sewer were such as adhered to the pipes (which were well flushed), there were frequent outbreaks of fever until 1866. Diseases which had formerly made their haunts in the lower parts of the town traveled by means of the sewers and infected the higher districts. In 1866 the sewers were systematically ventilated, and since that time there have been, until lately, no periodical outbreaks of fever, and, with a doubled population, " the rate of mortality rarely exceeds eighteen in the thousand, which is a standard of health unparalleled in the history of sanitary science for a district having so large a population." Quite recently a serious epidemic of typhoid made its appearance owing to defects which were described in Chapter II.

The principle of the ventilation of a sewer is practically the same as that adopted by builders for the prevention of dry-rot. The fungi which cause this rot in timber cannot produce their germs in a current of air, and if a sufficient number of ventilating openings are made, communicating with each other, the action of the wind from one side or the other will cause a sufficient current. So in a sewer, a continuous movement of the air in one direction or the other carries away and dilutes sewer gases, and if they contain germs of organic disease capable of infecting the human blood, these are believed to be destroyed by oxidation or otherwise.

A safe sewer always has a current of air passing through it, and if it contains sewage matters at all, these also must be in constant motion. On this incessant movement of the air and the liquid must we rely for our only security. A solution of sugar in water, remaining stagnant, and protected from a free circulation of air, will enter into a vinous fermentation. If well ventilated and agitated, no such fermentation takes place. It is asserted that the excrement of a typhoid patient, continually agitated in contact with fresh air and a fair admixture of water, passes through a series of complete chemical changes, with no injurious product; but if allowed to remain stagnant, if not freely exposed to the air, or if it gain access to human circulation before a certain oxidation, it will, like a ferment, reproduce itself, and give rise to the conditions under which it was itself produced. Motion and aeration are therefore needed to prevent infection, which is sure to be generated when typhoid evacuations are confined and stagnant. Unventilated and badly constructed sewers are sure agents for the propagation of the disease, when once it has taken root.

The resulting gases of sewer decomposition are the vehicle or medium for the conveyance of infection, and from their lightness they give rise to a rapid diffusion owing to the eagerness with which they seek means of escape at the higher parts of the sewer system, that is, in house drains, soil pipes, etc. It may not be possible entirely to prevent the development of the poison in even the best arranged sewer, but it is possible, by a free admission of air, to supply the oxygen which will take away its sting and render it harmless. Sewers which have large and frequent openings at the street surface, and through which the liquid contents have a constant flow, may give forth offensive smells, but, if they have proper attention, sanitary evils do not often result.

Sewer gas, when largely diluted on its escape (at frequent intervals) into the air of the street, is probably nearly or quite innoxious, but when it forces its way into the limited atmosphere of a closed living-room, the poison, or the germs of disease accompanying it, may easily work their fatal effects.

Sulphuretted hydrogen is found in all sewers in

which the sewage itself or the mucous matters adhering to the pipe assume a certain degree of putridity in the absence of a sufficient supply of fresh air. This gas is extremely poisonous; so much so that one part of the gas to two hundred and fifty parts of atmospheric air will kill a horse. At one half this intensity it will kill a dog. A rabbit was killed by having its body immersed in a bag of it, although its head was not inclosed and it could breathe pure air freely.

One of the most frequent sources of pressure upon the air within a sewer is the increase of temperature arising from the hot water escaping from kitchens and baths. The repeated expansions and contractions caused by the admission of hot and cold water produce a constant effect on all water traps connecting with unventilated sewers. With ventilation, the breathing in and out, as the air of the sewer contracts or expands, does not affect the water traps, because an easier passage is found through the ventilators.

The constantly changing volume of water in many sewers, as has been before stated, exerts a powerful influence on the confined air. As the water rises it reduces the air space, and if it reduces this to one half, it brings to bear upon the air a pressure equal to a column of water thirty-four feet in height, and this pressure is relieved by a forcing out of air through the most available channel, — the channel where there is the least resistance; if there is no other vent, a sufficient number of water traps must be forced to allow the pressure to become reduced. It being reduced, and the water falling again to a lower level, a vacuum is created which must be supplied by air forcing the traps in a reverse direction, and in either case the forced trap may remain open for the free passage of foul air until another use fills it with water. In the ebb and flow, too, a part of the perimeter of the sewer is made alternately wet and dry, with an accompanying production of vapor and gas.

As the chief domestic use of sewers is between morning and noon, and as at this time the most hot water passes into them, the pressure on the air in the sewer is during this period increased both by an elevation of the temperature and by a reduction of the air space. Then, from about noon until the next morning, the quantity of the flow decreases, the airspace increases, the temperature falls, and more air must be admitted to supply the partial vacuum created. Such fluctuations are constantly occurring, accompanied with a drawing in and forcing out of air, for which ample passage *must* be made independently of the water traps of houses, or sewer gas will surely enter them. Where proper air vents are provided, this ebb and flow of the sewer may be increased, with great advantage in the matter of ventilation, by artificial flushing arrangements which will allow the water to be dammed back and released at frequent intervals.

The movement of the air in and out of the sewer is also affected by barometric changes.

Where proper ventilation is furnished there will be an advantage in exposing the outlets of sewers to the direct action of the wind, but where there is not sufficient vent for escape, such outlets should, as has been stated, always be screened against strong currents of air.

Numerous experiments have been made with tall chimneys and fires, having for their purpose the creation of a strong draught from the sewer, but these are said in England never to have worked satisfactorily, and to be in no case recommended, being both expensive and troublesome. In Frankfort, on the contrary, such ventilating shafts are used with apparent good effect.

By reason of the causes constantly at work tending to the increase and decrease of the pressure of the air in the sewer, this variation may safely be depended on to furnish all needed ventilation, if only sufficient openings are provided from which air can pass in and out at frequent intervals.

Ventilation by rain-water pipes from the eaves of houses has often been recommended, but experience has shown that it is unsatisfactory, not only because it frequently discharges sewer gas near the windows of sleeping-rooms, but because at the time when ventilation is much needed these pipes are not available; either being filled with a rush of water or else having such a rapid downward current as to move the air toward the sewer rather than away from it, or because, from the position at which rain-water inlets are often introduced into sewers, these are entirely closed when there is a large amount of sewage flowing, — as during heavy rains, when ventilation is especially demanded.

This system was adopted during the early days of the Croydon work, and was rigorously pursued. In 1860 such ventilation was compulsory in all cases. The mortality was very much increased until a better system was adopted in 1866, when the death-rate fell again to its old standard.

In "Hints on House Drainage," by Dr. Carpenter of Croydon, we are told, with reference to fatal epidemics of typhoid fever, that the illness dated from two distinct times, at both of which, with a high temperature and a stifling atmosphere, there was a heavy fall of rain. "I do not mean to assert that each case commenced immediately after the rain-fall, but in upwards of twenty fatal cases into the history of which I examined, the commencement curiously ran up to two distinct dates, and of many slighter cases the patients stated that they had not felt well about the same periods." One case occurred in his own house. The water-pipe ventilators being closed by the rain water, and the air in the sewers being compressed by the increased volume of the flow, the gas forced the water trap of his soil pipe and escaped into his tank room, where the upper end of the ventilator was used as an overflow pipe for the cistern. This air ascended to a room occupied by two persons, both of whom were attacked with typhoid fever. There were no other cases in the house.

After all the experiments that have been tried with shafts, furnaces, mechanical blowers, steam jets, electricity, etc., the most experienced engineers have settled upon more frequent ventilation, by means of man-holes and lamp-holes opening at the centres of streets, as in all respects the best and safest. If these openings are sufficiently frequent and large, there is such an easy and thorough circulation of air in the sewer that the concentration of poisonous or of offensive gases is prevented, and their escape into the open air takes place at a point where they will be more diluted before reaching the sidewalks or the houses than if withdrawn by any other means yet devised. By the use of the charcoal ventilators described below, so arranged as to give free vent at their openings, it has been claimed that all practical danger or objection may be obviated. On the other hand, it is held by many competent authorities that even the best of these ventilators, while they do good as disinfecting agents, are objectionable as retarding the free circulation of air into and out of the sewer, which is the sovereign remedy for all the evil arising from the decomposition of the foul contents.

The great safety lies in the dilution of the gases by the free admission of air, and by their escape, when they escape at all, into the open air as far as possible from the house line. The effect of dilution is fully shown in fever hospitals; formerly, the mortality among both patients and attendants was frightful to contemplate; but now, although the ventilation is often far from complete, the condition of the patients themselves is much improved, and contagion is almost done away with; so much so that if an attendant contracts the disease it is taken as clear evidence that there has not been a sufficient dilution of the exhalations from the patients, or, in other words, that the ventilation has been imperfect.

The absorbing and disinfecting power of charcoal fully sustains its reputation. Latham quotes the following from Professor Musprat: " The absorbing powers of charcoal are so great that some have doubted whether it is really a disinfectant. This opinion has probably arisen from imperfect views of its *modus operandi*, since it not only imbibes and destroys all offensive exhalations and oxidizes many of the products of decomposition, but there is scarcely a reasonable ground of doubt remaining that it does really possess the property of a true disinfectant, acting by destroying those lethal compounds upon which infection depends."

Strictly speaking, the charcoal is simply an apparatus by which a natural process is carried on in an intensified form. It has the two important qualities of condensing upon the surfaces of its inner particles eight or ten times its volume of oxygen, and of attracting to itself all manner of other gases. It is not necessary that sewer gas be brought into direct contact with it by external pressure. By the operation of the law of the diffusion of gases, the impurities of the air next to the charcoal being absorbed, remoter impurities flow to this space and are in turn taken up, until the contents of a close room may be entirely purified by a small dish of charcoal. The oxygen that consumes or burns up the organic matter is speedily replaced from the atmosphere, and the constant efficiency of the apparatus is thus maintained.

The clogging of the pores of the charcoal with dust, or their saturation with water, prevents this action, and charcoal that has become wet or foul must be dried or burned in a retort before it becomes again perfect in its action. If charcoal ventilators are so situated as to keep dry and free from dust, they will not require changing or reburning more often than once a year.

The efficiency of even a small quantity of charcoal will be understood when we remember Liebig's statement, that a cubic inch of beech-wood charcoal contains a surface of interior particles equal to one hundred square feet. The especial adaptability of charcoal to use in sewer ventilators is further shown by the fact that it absorbs gases contained in or accompanied by the vapor of water (as they always

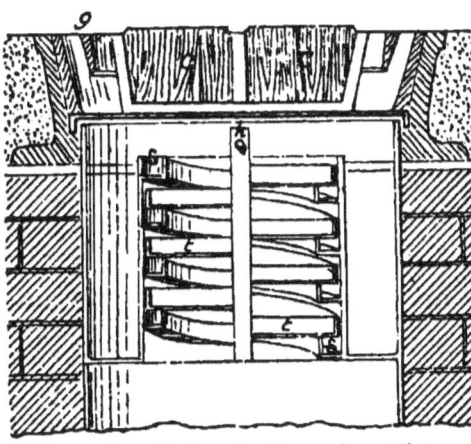

Figure 5. — Latham's charcoal ventilator for sewer and man-holes.

escape from the sewer) much more readily than those which are dry.

Several forms of charcoal ventilators have been devised. The best of them seems to be that of Mr. Baldwin Latham, which is a type of the class, all of which work on essentially the same principle. It is illustrated in the accompanying diagrams (Figs. 5, 6). The central cover, C, which is of wood, protects the charcoal from rain or water used in sprinkling the streets; g is a grating outside of the closed part, through which the air escapes from the sewer or is drawn into it. Under this grating is a dirt-box surrounding the ventilator and intended to catch dirt falling through the grating. There is an overflow (S) arranged to carry to the sewer all water reaching the dirt-boxes. 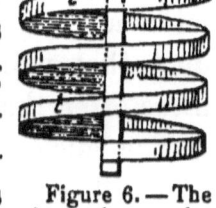 Figure 6.—The charcoal tray for Latham's ventilator.

The spiral tray t is made of galvanized wire-cloth and is filled with charcoal; it is screwed into the ventilator over the spiral trough S by means of the handle h.

The arrangement of this disinfector is such that all air escaping from the sewer must pass either through the charcoal or through the spiral passage between layers of charcoal. If the layers become so obstructed by dust that a free passage through them is not afforded for the air, there is still an easy vent through the spiral open spaces. The charcoal is thoroughly protected against dirt and wet, and will remain effective for a long time, and the arrangement is such that there can be no interruption of the working by the accumulations in the dirt-boxes, nor by the overflow of the water escaping from them. The sewer gas is all brought into close contact with charcoal, and has no possible means for escape except through the protected channels intended for it. The spiral tray should be filled with charcoal broken to about the size of marbles, and if care is taken in screening out its finer dust, it will afford a very permeable passage for gas. The dirt-box can be easily taken out and dumped, and readily replaced.

Ventilators should be closer together in the lower and filthier parts of a town than on higher lands or steeper inclines.

Mr. Latham thinks that they should never be more than two hundred yards apart. He advises renewing the charcoal once a month. Five hundred and sixty-two sets of his apparatus were used in Croydon. Their total cost, including labor, new charcoal, fuel for reburning, etc., made a charge of less than one dollar and twenty-five cents per annum for each. The charcoal is reburned in iron retorts having small pipes to carry away the escaping gases.

The usefulness of the charcoal ventilators is demonstrated by the fact that in Croydon the written complaints of smells from certain sewers coincided with the absence of the trays (taken out for repairs), and the cause of the complaint was removed y replacing them.

I am informed that notwithstanding the success

ARRANGING PLANS FOR TOWN SEWERAGE. 163

of these ventilators in Croydon, they are (there as elsewhere) being removed for the sake of giving a freer passage by simple open grating at the man-holes. It may therefore be assumed that however useful charcoal trays may be in cases where it is necessary to have ventilating openings close to houses, it will be best to dispense with them in street sewers, — trusting, rather, to the less obstructive communication through open gratings. (See note on page 171.)

On steep grades, where there would be a tendency for the air of the sewer to be drawn toward the ventilators on the highest land, discharging at this point an amount of gas that should be distributed along the whole street, it is well to place a light hanging valve in front of each outlet into a man-hole. Such a valve will not obstruct the flow of the sewage, while it will prevent the air below from finding its way up the drain, compelling it to escape at its own ventilator.

Where the ventilators used are not in connection with man-holes, they should rise, not from the crown of the sewer itself, but from a recess or chamber carried up to the height of a foot or more. Into this recess the sewer air will naturally rise instead of passing on up the line, as it would be likely to do were there only a small ventilator-opening to divert it.

With a free ventilation through the soil pipes at every house, there is an immense preponderance of area in favor of the vertical escapes, and these are frequently so placed that they become sufficiently heated to create a strong upward current. In a district containing a population of fifty thousand there would probably be ten thousand of these vertical openings, with a combined area equal to from twenty to forty times the area of the sewer at its mouth, so that their action would result more or less generally in the drawing in of air at the street openings ; a fact which is sufficiently proved in Croydon, by the accumulation of dust in dry weather in the charcoal-baskets with which the street openings are furnished. Where the orifice is a continuous exit, — that is, where there is no inward draught of air, — the charcoal remains black in spite of dusty streets.

It is a frequent practice with engineers to admit house drains at a very low point in the wall of the sewer, where they will ordinarily be entirely submerged. This renders such connections inoperative as a means for ventilating the sewer, and the ventilation of the soil pipes of houses so connected will consequently be of no avail as a part of the public system of ventilation. If the drain has no ingress for air at its lower end, the ventilation of the soil pipe itself will be much less complete ; the pent-up gases arising from the decomposition of the contained organic matters may escape, but there will be little of the needed circulation of air in the pipe. With a free sweep of air from below, this decomposition would not take place in a pent-up condition, but would be carried on with a full supply of constantly changing atmosphere. Under these circumstances the ventilation of the street sewer would have to depend upon its street openings alone. In a perfect system these should even play a somewhat secondary part, acting more as a means for the inlet of fresh air to supply the higher ventilators than as a means of escape for the air of the sewer itself.

All manner of chemicals used for disinfecting sewer gas are objectionable, from their unpleasant odor, their own injurious character, the constant attention their use demands, their inefficiency and their expense; nothing has yet been discovered that can at all compare with the simple use of wood charcoal.

THE FLUSHING AND CLEANSING OF SEWERS.

It is an important condition of all properly constructed sewers that they should be kept at all times entirely free from sedimentary deposit and from the adhesion of foul and slimy matters to their side walls. Theoretically, this cleanly condition should be absolute; practically, we must endeavor to approach as near to it as possible.

In practice, perfection in this respect is rarely to be attained throughout the whole series of sewers of a town. It will more or less often become necessary to resort to some artificial means for removing sedimentary deposits, and for washing the walls of sewers which have become coated with slime, — the deposition of which is one of the evils to be guarded against on the score of health.

If from imperfect construction, from too great

size, from lack of flow, or from any other cause, the sewers do not keep themselves clean with their natural flow, then it becomes absolutely necessary that their deposits be removed either by hand labor or by flushing with copious floods of water. The less this is needed, of course the better, but if needed it is quite imperative.

The purpose of all flushing is to create a flow of sufficient depth to be thoroughly cleansing, and to keep it up for a long enough time to wash away all accumulated matters.

For small sewers, where an intermittent sudden addition of a few cubic feet of water to the natural flow would be sufficient to keep the line clean, a tumbler tank set on trunions directly in the line of the sewer and in connection with the man-hole will be useful. The accompanying illustration (Figure 7) of such an apparatus is taken from Latham's "Sanitary Engineering," where it is thus described:

" When empty, this tank would remain level, as the portion B behind the trunion is heavier than the portion A before the trunion, but when the tank fills with water or sewage the portion A becomes the heaviest, and the consequence is the tank tilts, discharging its contents into the sewer below, and afterwards righting itself ready to receive a fresh charge."

The tank should be made of cast iron, and its end, as it falls in either direction, should be received on wood, rather than on stone or metal.

Special flushing appliances are frequently required. These appliances are Dams; Reservoirs; Tidal Basins; the use of Street Hydrants, etc. Each of these may have its special advantages for special circumstances.

Dams are obstructions placed in the sewers for holding back to a greater or less extent the sewage coming from above until its volume shall become

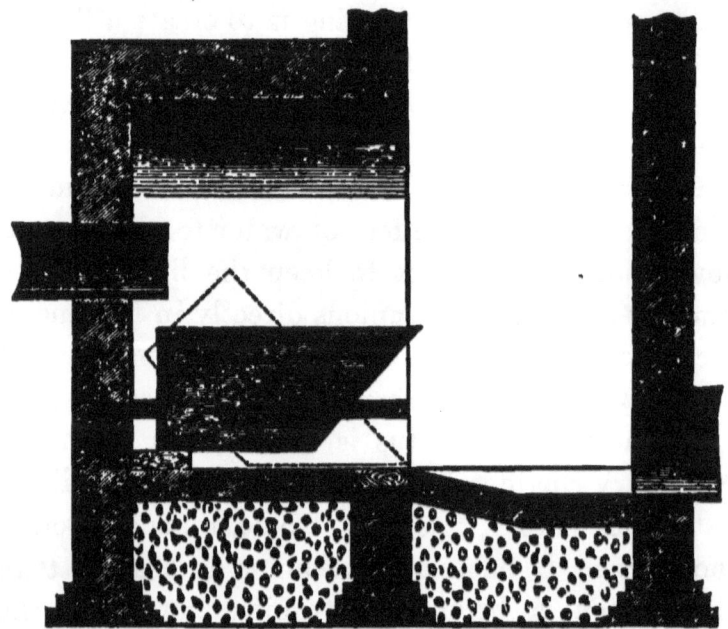

Figure 7.

sufficient (on the removal of the dam) to wash clean all that part of the line lying below them. Dams may be either elaborate cast or wrought iron appliances (sometimes with gun-metal facings) or any cheaper, or simpler contrivance, down to a simple wooden gate to be raised or lowered from the mouth of a man-hole. The dam may close the sewer entirely, or only partially. Where the grade is slight so that the damming back of the water to half the diameter of the sewer will set it back for a long distance, giving in this way sufficient volume, a half-dam will suffice, and it has the advantage, that if forgotten or neglected the flow of the sewer will not be totally obstructed, there being sufficient

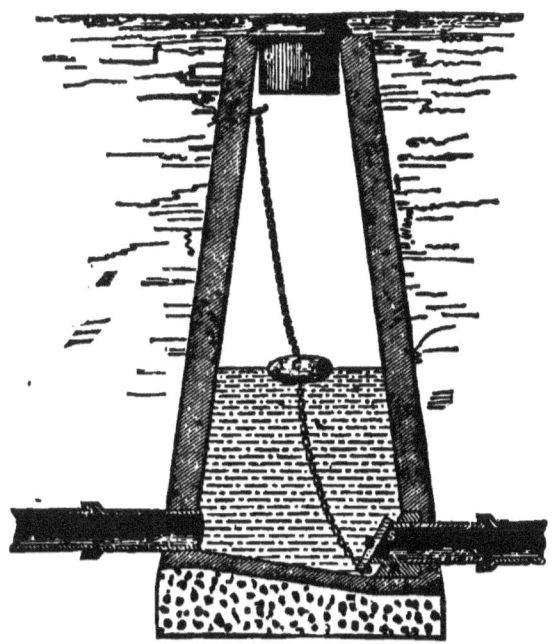

Figure 8.

water way left over the top of the dam. Where it is necessary to choke the flow entirely and to close the whole sewer with a dam, a special overflow way should be provided, in order that if this dam is not removed in time, there can still be an escape for the accumulation from above to avoid the danger that

ARRANGING PLANS FOR TOWN SEWERAGE.

the sewage will set back into the house drains. The modifications of the system of damming and the variety of apparatus for preventing the set back of the dammed sewage into lateral branches or house drains are so various, and the requirements for them differ so much with different circumstances, that it is hardly worth while to describe them minutely here.

As an illustration of such apparatus, the pipe sewer dam in connection with man-holes, shown in Figure 8, is taken from Latham, who thus describes its action : —

"For small sewers the author has used an earthen-ware flushing block which is built into the head of every sewer running out of a man-hole as shown at A. These flushing blocks have a ground face against which a wooden disk, B, is placed. The presence of the water tends to fix the disk in its position, and the disk is connected to a chain. To guard against neglect, the float c is fixed on the chain, so that if the disk is left fixed in the sewer or as soon as the man-hole fills with sewage to such an extent that the float begins to swim by its power of floatation, it liberates the wooden disk from the mouth of the sewer, and the sewage escapes to the lower level."

Reservoirs are frequently useful for the accumulation of either sewage or extraneous water in greater or less quantities, to be ultimately set free to flood the sewers. Frequently the water of a brook or a storm flow may be used to supply these reservoirs, — which have the advantage that their contents may be retained for any desired time, and let loose whenever flushing is most required.

Tidal basins (or tidal reservoirs) are receptacles to be filled by the high tide, and let loose at low tide, and are chiefly useful at the upper ends of long flat grades emptying into tide water. By their means even absolutely level sewers may be kept quite clean from deposit.

Street hydrants are useful for large sewers when used in combination with dams or reservoirs, and for smaller lateral sewers by their direct flow.

The arrangement of these flushing appliances, one or more, or all of them being useful according to circumstances, must of course depend upon the conditions obtaining in different localities. There are few towns where some or all of them are not needed for the comparatively satisfactory working of the sewers.

In New York, in 1867, the cost of cleansing about two hundred miles of pipe sewers was only one hundred and twenty-five dollars.

The following extract from a statement made by the engineer for the construction of sewers of the city of New York (published February 3, 1873), illustrates the comparative foulness of brick sewers and pipe sewers: —

" The usual price of cleaning sewers by hand is about $2.50 per load, and while under good sewer system solid deposits should be carried off with the flow, the city has been yearly paying from

$27,000 to $46,000 per year to remove them. It is notorious that persons who, under the old Tammany *regime*, had contracts for cleaning these sewers, finding it profitable to remove the deposits at $2.50 per load, were in the habit of putting obstructions in the sewers with a view of creating solid deposits. The present commissioner of public works has, however, put a stop to all this, and last year reduced the cost of cleaning the sewers to $14,412, against $44,690 for the year 1871. The following table, furnished by Engineer Towle, shows the comparative cost of cleaning brick and pipe sewers from 1867 to 1871 inclusive. The water supply having increased last year, the department has resorted to the flushing process, and two or three nights per week the water from the hydrants has been let into the sewers, reducing the expense of cleaning for the year to $14,000.

Note to page 150. — By the use of Field's Sewer Flushing Tank, it is no longer necessary to rely on rain-fall for this work.

Note to page 163. — More recent experience has clearly shown the advantage, save in special cases, of depending exclusively on open ventilation at the man-hole, rather than to obstruct the flow of air by the interposition of charcoal screens.

| Years. | BRICK SEWERS. | | | PIPE SEWERS. | | | Proportion of Brick to Pipe. | Proportionate Cost of Cleaning Brick to Pipe Sewers. |
	Total Length in City in Linear Feet.	Number of Loads Removed.	Cost of Removal.	Total Length in City in Linear Feet.	Number of Loads Removed.	Cost of Removal.		
*1867	1,058,136	13,073	$32,682	150,022	50	$125	7 5-100 to 1	261 46-100 to 1
1868	1,068,817	19,358	48,295	222,020	80	200	4 81-100 to 1	241 48-100 to 1
1869	1,088,911	11,002	27,730	288,120	200	500	3 78-100 to 1	55 46-100 to 1
1870	1,120,234	18,548	46,420	335,313	597	1,442	3 34-100 to 1	32 18-100 to 1
1871	1,152,054	17,374	43,435	346,903	502	1,255	3 32-100 to 1	34 61-100 to 1

* The construction of pipe sewers was commenced about 1865, and no considerable amount was paid for cleaning them until 1869.

The cost given in the schedule for pipe sewers includes the expenses of repair, the removal of broken pipes when encountered, and relaying new ones.

In all cases where the pipe sewers have required cleaning or repair, their failure to work has been traced to error or unfaithfulness in their construction.

It was formerly supposed that with pipe sewers not too large for the amount of liquid they were to carry, there would be no necessity for flushing, and so far as sedimentary deposits are concerned this is usually true; but a slimy coating often forms on the wall of the pipe and enters into decomposition, generating objectionable sewer gases. For this reason, all pipes used for house drainage only should be so arranged that they can be occasionally flushed out with a good flow of fresh water; but where rainfall is admitted from roadways and from the roofs of houses, additional flushing will not, generally, be needed, except during epidemics, or in dry, hot seasons. At such times there is always a great advantage in frequent flushing, and occasional disinfection.

In flushing, always begin with the lower part of the system, nearest to the outlet, and work back toward the heads of the lines, so that there shall be no danger that deposit already existing in the lower parts will stop that coming from above in such a way as to cause the complete choking of the channel.

Hand cleansing is to be avoided whenever possible, and the circumstances are few under which it is necessary to construct a system of sewers in which this costly and objectionable process will be required. It is now and then inevitable.

STREET GULLIES.

So far as it is a part of the plan to take surface rain-water into the sewers, proper openings must be made at street corners, or elsewhere, according to the rapidity of the inclination of the gutter, the location of the lowest point of the grade, the extent of surface to be drained, the frequency with which it is desirable to establish flushing points, and other considerations which may arise under certain circumstances.

These street openings should not be used as sewer ventilators, that is to say, that they should be well trapped in order to prevent the air of the sewer from escaping at the side of the foot-way. In addition to their trapping they should be provided with some form of receptacle for the arresting of the sand and other heavy detritus washed from the roadway.

Various devices have been adopted to secure the admission of surface water from gutters to the sewer without allowing the escape of sewer gas. These are usually arranged with a deep recess below the outlet for the accumulation of sand and silt washed from the roadway, and with some form of water trap. Their construction in our northern climate should have careful reference to a severe action of the frost, and

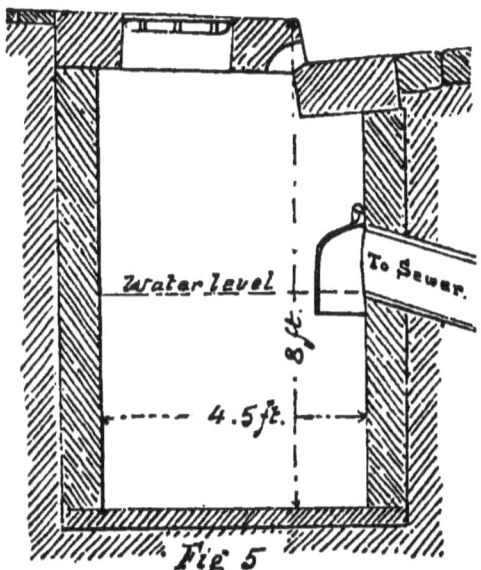

Figure 9. — Catch-basin for admitting street wash.

no plan that has come under my notice seems so well adapted for this as one used by Mr. J. Herbert Shedd, the engineer of the sewerage in the city of Providence, the arrangement of which is shown in the accompanying diagrams. The trap for sealing

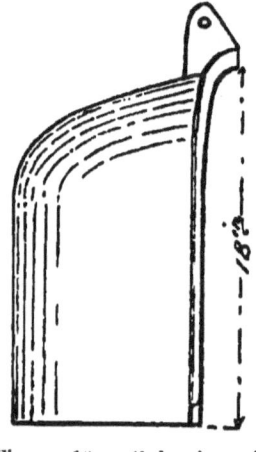 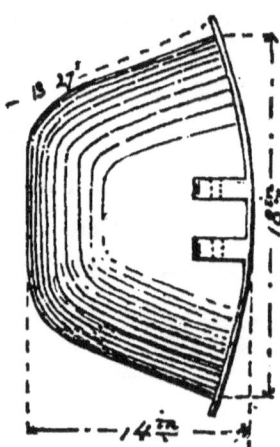

Figure 10. — Side view of catch-basin trap.   Figure 11. — Top view of catch-basin trap.

the outlet is made of cast-iron, hinged with a copper bolt. It is firmly attached to the side of the basin with cement, and, if disturbed by frost, is simply torn loose from the brick-work, and can be easily cemented to its place in the spring.

MAN-HOLES AND LAMP-HOLES.

All sewers should be provided with man-holes for ventilation and for service during examination; and pipe drains should have, between the man-holes, and at every point where the vertical or horizontal direction of the sewer is changed, lamp-holes, at the bottom of which lanterns may be suspended which will enable the line to be examined from the nearest man-hole. The removal of all such obstructions accumulating in pipe drains as cannot be washed out by flushing is effected by various instruments attached to jointed rods, like chimney-sweep tools, which serve as handles, enabling them to be used even at a distance of several hundred feet.

The man-hole is a shaft or chimney built up from the sewer to the surface of the street, having an opening large enough for men to enter for work and some provision for steps to enable them to descend easily.

Lamp-holes are smaller shafts which may be furnished with vitrified pipe reaching to the surface of the street, with an opening sufficiently large to allow a lantern to be lowered into the sewer. As all sewers should be straight (vertically and laterally), between all man-holes and lamp-holes (except in rounding corners), by sighting from a man-hole to a lamp-hole, the workman can easily determine whether the sewer is obstructed or free.

Man-holes and lamp-holes are generally covered with tight cast-iron caps, but it would be better in all cases to substitute for these strong iron gratings with the largest possible air spaces, the cover being raised very slightly above the grade of the street in order that there shall be no surface flow into it. And, even when this is done, it is often advisable to make a recess or catch-pit in the bottom of the sewer under the man-hole, to retain any earth entering through the grating. The advantage of the grating over the close cover is, that it gives the best means of ventilating the sewer that has yet been devised.

PRIVATE DRAINS.

The public sewer or drain may properly afford an outlet to the land drainage of private property, but before reaching the public drain, this should pass through at least two rods of sub-main drain laid under the direction of the public engineer, and trapped as he may direct for the exclusion of silt or refuse. This sub-main should deliver its water into the public drain as nearly as possible in the direction of the flow of the latter, so that the streams may run together without confusion, and the danger from eddies be obviated. Drains from houses and all private establishments should be connected with the sewer under similar official regulation, and should enter at sufficient height to act as ventilators of the sewer, — their own ventilation being aided by this means for the entrance of air from below.

SEWER JUNCTIONS.

The character of the junctions of main and tributary sewers has much influence on their capacity. It has been found that when equal quantities of water were running in two sewers, each in a direct line, at a rate of ninety seconds, if their junction was at right angles their discharge was effected only in one hundred and forty seconds, while if it met with a gentle curve, the discharge was effected in one hundred seconds.

In one recorded instance, a pipe, having been gorged by reason of a right-angled junction, which kept the velocity of its flow down to one hundred and twenty-two feet per minute, had its flow increased to two hundred and eight feet per minute, and the difficulty entirely removed by making the junction on a curve of sixty feet radius. The same objection holds with right-angled junctions falling vertically into the sewer. In this case, as in the other, the inlet should be on a curved line ; but vertical junctions are usually objectionable.

Frequent junctions are of great advantage. Experiment has shown that, with a pipe having a fall of one in sixty, its capacity, with junctions at frequent intervals, is more than three times what it would be if flowing only from a full head at the upper end of the pipe. In sewers of larger sizes the capacity is increased more than eight times.

TIDE VALVES.

In all cases where the outlet of the sewer is the level of tide water, it is considered desirable to establish some form of valve or dam which shall prevent the tide from setting back into the sewer.

For large sewers, the most available form of tide gate seems to be that used in the case of canal locks, — two gates swinging together by the pressure from without, and holding back the rising waters ; opening as soon as these waters recede below the

ARRANGING PLANS FOR TOWN SEWERAGE.

level of the sewage within. All discharging of sewage below the level of high water is to be avoided whenever possible, unless for the removal of storm waters; for although the ebb and flow of the tide in a sewer suffices to keep the part thus traversed clean, there is usually a sedimentary accumulation about the point reached by the top of the tide water flow. So far as the ordinary house sewage is concerned, it should, even if pumping is necessary, be kept constantly in motion, except in those cases where it is desired to store it temporarily for flushing purposes.

CHAPTER V.

THE CONSTRUCTION OF SEWERS.

IN a work not intended for the practical guidance of engineers, it is not necessary to say very much on this head beyond what is contained in the preceding chapters. In determining the size of sewers needed for the area to be drained, — making allowance for the inclination or steepness of the grade on which the line is to be made, and for the proportion existing between the regular flow of sewage and the amount of storm water that it will be necessary to provide for, — the considerations heretofore set forth, will be a sufficient guide to any engineer qualified to take the direction of such work. To give all the necessary details required for the instruction of a novice would be far beyond the scope of this work, nor would it be possible here to qualify amateurs, and sewer authorities who lack professional experience, to take the direction of the work into their own hands.

My chief purpose being to show to the private citizen or to the average chairman of a committee of aldermen, or supervisors, what are the essential requirements of good sewers and what they must demand in order to secure the best conditions

THE CONSTRUCTION OF SEWERS. 181

of health, rather than to instruct them how to carry out the technical work of construction, it is not necessary to say more in this chapter than is needed to call the attention of such persons to matters which are not always sufficiently regarded by the class of engineers who are employed for local works of sewerage.

The most essential condition to be sought in all work of this class is *the utmost possible thoroughness*. Without wishing in any way to reflect on the character of the management of sewer construction in the country generally, and having no doubt that there are many engineers who manage the details of their work with equal care, I would seriously advise the authorities of any town where the organization of a complete system of sewerage is contemplated, to visit the works now being carried on in Providence, Rhode Island, under the very thorough management of Mr. J. Herbert Shedd. One may learn here, better than in any other place with which I am familiar, the real meaning of the word "thoroughness" as applied to sewer construction.

The city supplies all the material used, selling these at established rates to the contractors, and it then devotes the energies of competent inspectors to securing the nearest approach to perfection that is possible in every branch of the supply department. All cement bought is subject to the condition that, after having been properly mixed with as little water as practicable and exposed half an hour to the air and then immersed twenty-four hours in water, it shall stand a traction strain of sixty pounds to the square inch. Every barrel received at the dock is numbered, sampled, and tested, and not one is allowed to be used which breaks with less than this strain. Every barrel not coming fully up to the standard is thrown on the hands of the parties supplying it. Sewer pipes are required to be of the requisite thickness and hardness, and practically true in form; if pipes are delivered which are defective in any one of these respects, they are rejected. All bricks bought are to have a certain high average of quality and form, and to contain a due proportion of extra hard specimens, and none too soft for the best work. If the lot is defective, when measured by this standard, the manufacturer must seek another market for them. Accepted lots of brick are sorted into different qualities according to form and hardness, the very hardest being used for the bottom of the sewer where the greatest friction is to be resisted, and the less hard ones (none approach softness) for the upper parts of the work. Mr. Shedd frequently uses tubular inverts made of earthen-ware like sewer pipes. A chief objection generally urged against these is that their form is so much affected by warping in manufacture that they do not constitute a good foundation for true brick work, but at the Providence depot only the really perfect ones are accepted. And it is no doubt very often from what has been rejected there that the usual supply of the market has to be drawn.

It is undoubtedly, to a certain extent, to the disadvantage of other places that the Providence rejections are so copious, for what is there discarded is not wasted, — only sold to those who are less particular. But it would be decidedly to the advantage of any town undertaking works of this sort to place themselves at once in the category of those who insist on having the very best material or none.

I have heard contractors who are accustomed to the slap-dash manner of sewer building, that prevails over the country generally, complain bitterly that the Providence engineers and inspectors are so rigid concerning every detail of the work that a contract undertaken there is very apt not to be profitable. At the same time there seems to be no lack of contractors, who are willing to do the work of this city in the manner demanded of them, and the result has been no doubt as nearly perfect as any sewage work in this country.

While much of the quality of any public work is due to the chief engineer having it in charge, and to the regulations that he establishes for securing good material and good workmanship, an even more important duty falls upon the inspectors, for it is they who are to watch the quality of every item of the material and the character of every foot of the work. However good the regulations which they are charged with carrying out, these regulations will be of but little value unless the carrying out is thorough and conscientious, and is entirely uninfluenced by the seductive efforts of contractors to secure their neglect of duty or obliquity of vision.

As the weakest link of a chain is the measure of its strength, so is the weakest part of a drain or sewer the measure of its permanent usefulness. It does not suffice that the work is on the whole good and reliable,— it must be good and reliable *in every part*.

Many pages might be written concerning the details of the art of sewer making, but I propose to leave the whole of this part of the subject to the technical works in which it has already been so well treated — except to say that there is in this country a quite universal tendency to excessive expenditure in the thickness of the walls of sewers. When the brick and cement are of good quality, and when a proper natural, or artificial foundation is secured, it is not necessary to make the wall thicker than one ninth of the interior diameter of the sewer, or as nearly this as the thickness of the material will allow. A single course of brick (four inches) is ample for the wall of a three foot sewer, but if the sewer is made larger than this, it will be necessary to increase the thickness to eight inches up to a diameter of six feet, and to add another course of brick if the sewer is larger than six feet.

The main sewer in Saratoga, nearly two miles long, and traversing ground of very varying quality, has a uniform diameter of three feet, and is made throughout its whole length (except at one point where unusual pressure may, under certain circumstances arise), only four inches thick. An inspection made after the completion of the work shows

THE CONSTRUCTION OF SEWERS.

an absolute continuity of wall and regularity of form, save in one short stretch where, from the contractor's neglect of properly sheet-piling quicksand, there is a slight deflection, not enough, however, to require the re-laying of the sewer.

All who are interested in the control of sewage works should adopt it as their leading principle to secure a well considered and suitable plan; to intrust its execution to a conscientious and competent local engineer; and to secure inspectors who will make it practically certain that every department shall be thoroughly and faithfully executed, in spite of the skillful efforts of old contractors to secure an opportunity to "scamp" their work.

CHAPTER VI.

THE DETAILS OF HOUSE DRAINING.

THE various items of the work of draining the house concern both the architect and the engineer. The latter, in so far as it relates to the admission into the public works of sewerage of the liquid refuse of the house, and the making of the necessary provisions to prevent any injury being done to the public interest by reason of careless or improper connection or the admission of improper substances; the former from the still more important considerations connected with the proper arrangement of the house as a domicile for human beings.

The architect should concern himself especially with all that relates to disposition of the necessary wastes and offscourings of domestic life. Whether it be a question only of disposing of the kitchen wastes or whether the most complete plumbing appliances are to be introduced, every part of the work should be so planned as to accomplish his purpose completely, entirely, and permanently, with due precaution against the entailing upon the occupants of the house of the evil results that a badly arranged system will be sure to produce.

It would be impossible in the short space of this

THE DETAILS OF HOUSE DRAINING. 187

chapter to give in detail all that may be needed under the great variety of circumstances arising in a varied practice. All that can be hoped, is so to set forth the general principles by which the work should be guided, the specification of what is to be avoided, and some of the more usual processes which are to be recommended, as to enable the person having charge of the building to apply his own judgment and discretion in the matter in such a way as to secure the most satisfactory end. Much of what is contained in the first three chapters of this work have a direct bearing upon the details of house drainage and whatever is here said is to be read in the light of the information therein given.

PLUMBING ARRANGEMENTS, ETC.

The accompanying diagram shows the simplest form in which the plumbing and draining of a house can be arranged to render it absolutely safe. An important feature of the plan here shown is that of providing a separate reservoir of water for the supply of each water-closet; this, though not unusual, is far from universal, and it is the only efficient means for preventing the tainting of the main water-supply pipe of the house with the gases formed in the basins, and the sucking into the main of the foul air above the trap when the water falls away in the pipes, as from the opening of cocks in the lower part of the house.

Referring to the diagram (Figure 12), which shows the general arrangement of plumbing, etc., it is to be

See reference to this cut in Chapter XII. The arrangement here given is manifestly bad.

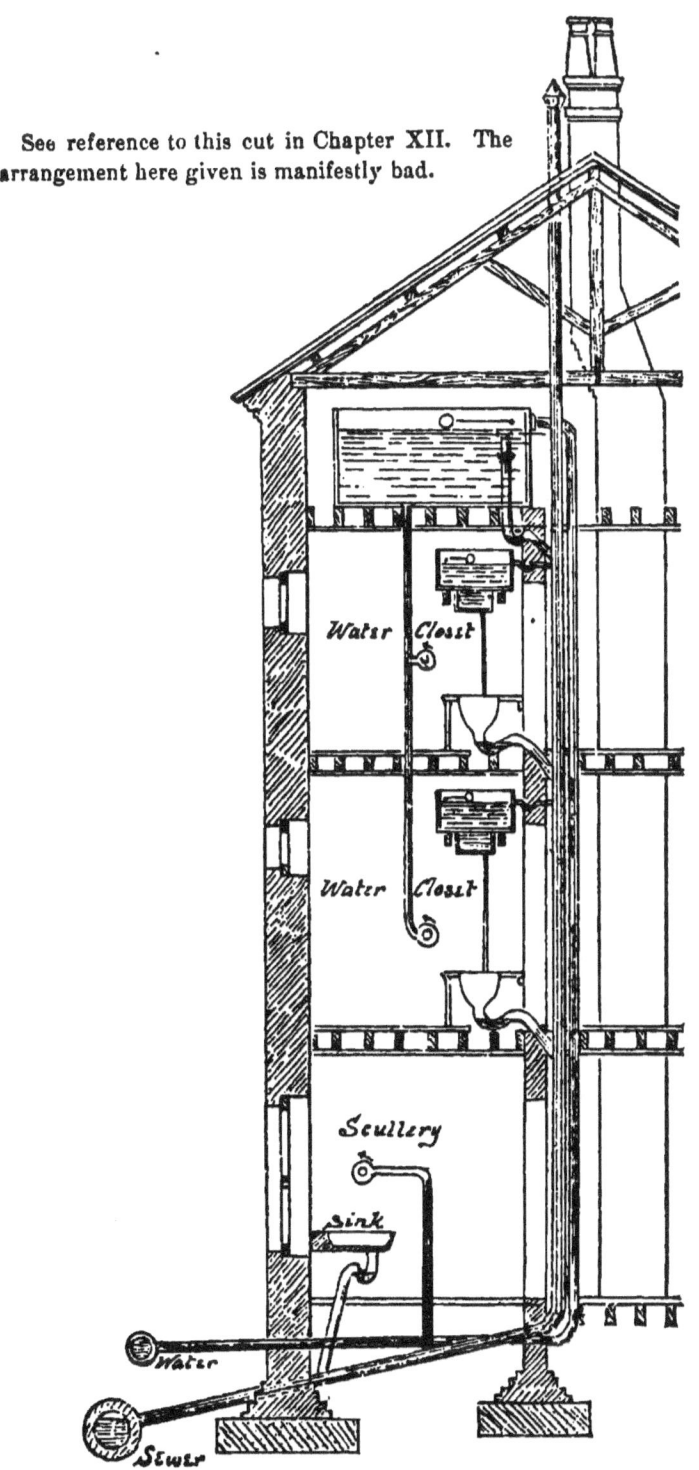

THE DETAILS OF HOUSE DRAINING. 189

said that from a sanitary point of view the most important feature there shown is a complete ventilation of the drain leading to the sewer, so that by no possibility can there be a forcing back into the house of gases formed in the sewer or in the main drain. As already stated, a usual water trap, no matter how deep, does not suffice to secure this. A water trap having a bend of even two feet would resist a pressure of only about one pound to the square inch, while the sudden filling of the sewer, by rising tide or falling rain, to such an extent as to reduce its air space one half, would bring to bear a pressure of fifteen pounds to the square inch; and whether the filling be sudden or gradual, the degree to which the increased pressure would affect any given outlet would depend on the facilities offered elsewhere for the air to find vent. In our ordinary town sewerage works, it is never safe for the householder to depend on other vents than his own connecting drain being available; he must in self-defense assume that his own drain is the only channel of escape, and make it impossible that air escaping there should find its way into the house.

Where severe frosts are not to be guarded against, this may be accomplished by discharging the water of the house into a receptable that is open at its surface, and from which a drain passes to the sewer with some form of trap; into this surface opening, for greater cleanliness, a rain-water pipe from the roof should discharge. Under this arrangement, if sewer gas is forced from the drain it will escape into the outer air. The chief objection to the plan lies in the fact that such escape would too often take place where it would be offensive, and sometimes too near an important window. A much better plan is to furnish a fresh-air inlet at the lower end of the soil pipe or house drain which will supply a purifying current of air through the whole length of the soil pipe. Some form of covered grease-trap, or flush-tank may be used, with a ventilator not less than two inches in diameter, and by the straightest available course, from this to a point well above the highest dormer windows. An opening should be made for fresh air in the cover of the grease-trap.

WATER-CLOSETS.

Dr. Simon, in his report of 1874, states the following as imperative conditions that should be insisted on wherever water-closets are allowed: —

"1. That the closets will universally receive an unfailing sufficiency of water properly supplied to them.

"2. That the comparatively large volume of sewerage that the system produces can be in all respects satisfactorily disposed of.

"3. That on all premises which the system brings into connection with the common sewers, the construction and keeping of the closets and other drainage relations will be subject to skilled direction and control."

In his explanatory remarks he states : That a sufficient supply of water is a supply that will enable each closet to be well flushed whenever used, and that the supply must be not only professedly, but actually constant. The best way to secure this is to supply to each closet from an independent cistern immediately above it. That every privy drain must be properly trapped and ventilated, and properly constructed, — ventilation of the soil pipe above the roof being imperative. That wherever practicable the connection between the house drain and the sewer should be through a trapped, open gully covered with a grating. He considers the *ordinary* water-closet thoroughly unreliable for those who are unlikely to take proper care of it, or who are too poor to keep it in repair, — no form of indoor privy should be allowed for this class, and even in the best houses water-closets should never be so placed that they cannot have outside windows.

For classes from whom the ordinary water-closet should be withheld, some suitable form of water-closet specially constructed for them, and constantly superintended by the public authorities, seems to be under the proper circumstances the best convenience yet devised.

The water-closet should never be supplied direct from the water main, but always from a separate reservoir, so that there shall be no danger of the sucking back of the contents of the pan when the water falls, as it so frequently does, from the supply pipes.

Dr. Hill, the medical officer of health for Birmingham, in a paper on sanitary improvements says: "What I wish to bring out prominently is, that water-closets being in direct communication with the sewers, which they imperfectly close when the valve is at rest, and actually open when it is in action, and being placed in the interior of houses, must inevitably be the means of introducing poisonous sewer-gas into dwellings, and so act as a source of danger and injury. The question then, arises, how are their ill-effects to be guarded against? I think best in one of two ways (traps, it is admitted, are such in more senses than one, and are useless); either by having them quite detached from the house, or partially so by double doors and intervening lobby, with good cross ventilation."

Dr. De Chaumont says, "Under no circumstances ought there to be a closet opening directly into a bedroom, the merely occasional convenience of such an arrangement being more than counterbalanced by its danger and generally objectionable situation. In some houses, however, particularly in older houses in towns, I have seen this arrangement, not only in single instances, but in all the closets of the house, so that access to one was only obtainable through a bed-room. Almost equally objectionable is the arrangement where the closet opens on to the lobby or landing close to the bedrooms or sitting-rooms, a plan both unhealthy and in every way offensive."

The usual "pan" closet is in several ways objectionable; chiefly as containing in the chamber

THE DETAILS OF HOUSE DRAINING. 193

beneath the pan a certain quantity of fouled water above which is an unventilated air space, sometimes, from imperfect construction, leaking its gases into the room, and always sending up a fœtid whiff when the pan is tipped.

The Jennings closet, shown herewith (Figure 13), has the peculiarity that it contains directly under the seat the whole charge of water to be used for the flushing at each operation of the closet. Fæcal matters are immediately immersed and so at once somewhat disinfected, and on the lifting of the valve the whole volume is rapidly carried away through the water trap into the soil pipe. The whole apparatus, from the seat to the soil pipe, is a single piece of earthen-ware, and the valve is held so firmly in its place by its own weight and by that of the water bearing upon

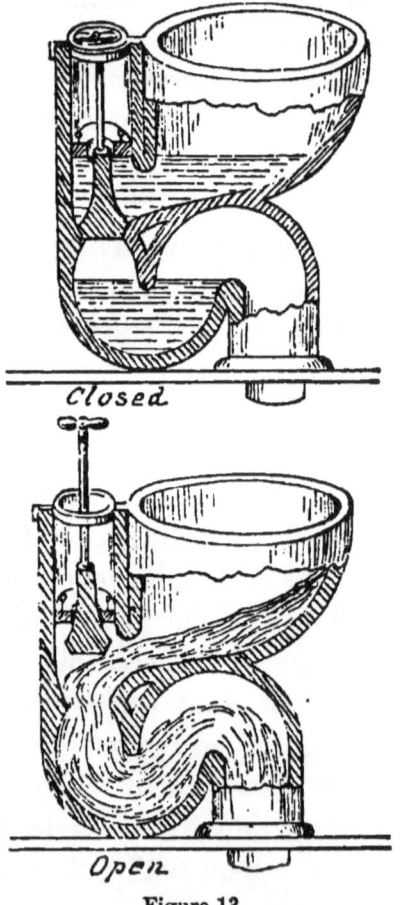

Figure 13.

it, that *if proper vent is given to the soil pipe itself,* so that the pressure of sewer air cannot be brought to bear upon it, and so that its trap cannot be siphoned out, there is no probability of the least escape into the room. If the soil pipe is not ventilated, the Jennings closet is the worst of all. In any case it should always be supplied with some arrangement like Blunt's siphon-overflow cap. (See chapter XII.)

SOIL-PIPE VENTILATORS.

As is sufficiently explained in other sections of this book, the cardinal remedy for the sanitary evils arising from the invasion of houses by the poisonous gases of sewers and drains, lies in the thorough ventilation of soil pipes by pipes opening through the roof into the free air.

Such ventilating pipes should be made of some permanent material. Earthen-ware is objectionable, lead and cast-iron are good and reasonably durable. Zinc — and consequently the zinc coating of galvanized iron — is very subject to decay under the action of the corrosive gases issuing from soil pipes. When galvanized iron pipes are used, they should be thickly coated with paint on their insides. If a free current of air passes constantly through such a pipe, — taken in from an opening in the waste pipe or catch basin outside of the house, and discharged above the roof by a large pipe, — the formation of corrosive gases will be much reduced. In northern latitudes the effect of frost must be guarded against.

It is especially important that soil pipe ventilators should be as nearly straight and vertical as possible; a crooked ventilator pipe will not "draw" any more than will a badly built chimney flue, nor even so well, as it lacks the heat of a fire to set up a current.

GREASE TRAPS.

There are various forms of grease trap which serve a good and useful purpose. The best that has come to my notice is that shown in the accompanying diagram.

Figure 14.

It is made of well-cemented brick-work, and need not be more than from four to six feet in diameter (according to the liberality with which water is to be used in the house. It must be absolutely water-tight. It should be placed close to the house, so that there shall be the least practicable length of drain pipe to accumulate grease, — allowing it to flow hot into the trap, where it will float at the surface of the liquid, at a point at least a foot above the mouth of the bent outlet pipe. Any solid refuse will have ample room at the bottom of the trap, well below the outlet. I have not found it necessary to clean out my trap (made in this way) more often than once a year. Indeed, the solid deposit, being organic matter, decomposes and forms ammonia, which helps to dissolve the grease and make it soluble, so that both the deposit and the scum are constantly being washed away. It would be well to run a rain-water spout into this trap to help cleanse it, and if it is not near a window, its best ventilation would be by a grating in its top-stone.

THE DISPOSAL OF HOUSE SLOPS.

Obviously, no form of grease trap or tight cesspool can serve for the final disposal of house slops. It is only an intermediary step in a process whose further course it is very important to direct. The treatment of this matter has been so successful at my own house, that the system there in use seems worth describing in this connection.

The house drainage is discharged into a tightly cemented tank four feet deep and four feet in diameter, entering near its top, which is arched over and closed by a tightly fitting stone cap, and thoroughly ventilated. This tank is similar to that described above. Its outlet pipe, starting from a point one foot below the surface of the water and about two feet below the cap-stone, passes out near the surface of the ground and is continued by a cemented vitrified pipe to a point about twenty-five feet farther away. Here it connects with a system of open-jointed drain tiles, consisting of one main

THE DETAILS OF HOUSE DRAINING. 197

fifty feet long, and ten lateral drains six feet apart and each about twenty feet long. These drains underlie a part of the lawn, and are only about ten inches below the surface. During the whole growing season their course is very distinctly marked by the rank growth of grass over and near

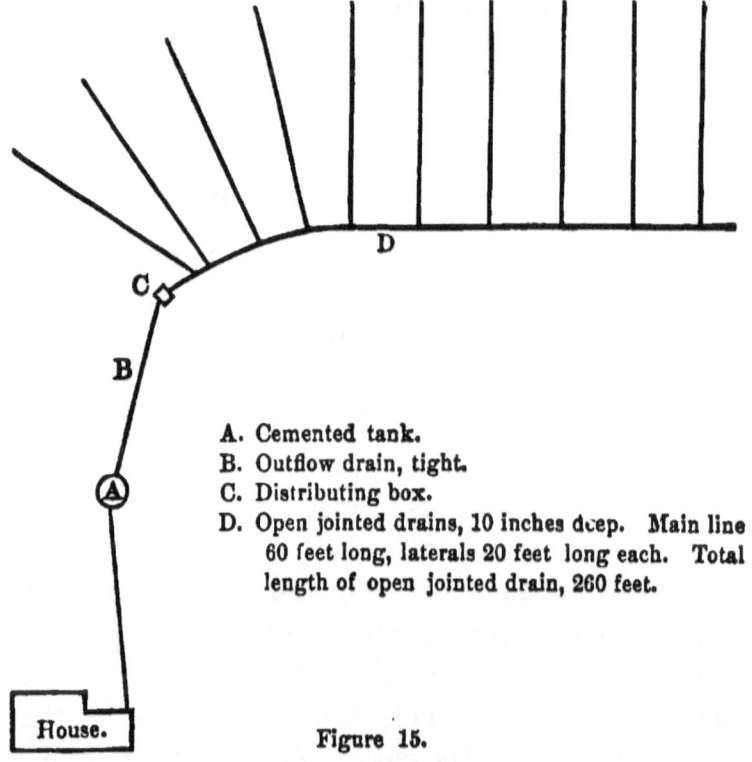

A. Cemented tank.
B. Outflow drain, tight.
C. Distributing box.
D. Open jointed drains, 10 inches deep. Main line 60 feet long, laterals 20 feet long each. Total length of open jointed drain, 260 feet.

Figure 15.

to them, the difference of growth in their immediate vicinity being so great that were the work to be done over again, I should place the lines but three feet apart. The slope of the ground is very slight, probably not more than fifteen inches between the extreme ends of the system, yet, judging by the growth, the distribution is very uniform through all the pipes, — main and laterals.

The arrangement of these drains is shown in Figure 15, an arrangement suited to the conditions of the place, but sufficiently illustrating the general principle.

I supposed, when I first adopted Mr. Moule's suggestion to make this disposition of the house sewage, that some other arrangement would be necessary, for the winter season, but even during the winter of 1874-75, — the coldest for many a long year, — the liquid has been perfectly disposed of, and has apparently found its outlets equally in all parts of the drainage. (See chapter XII.)

Successful though this experiment has been, I have recently adopted a small Field's flush-tank, in the belief that the system would be improved by having the discharge made intermittent, so that the flow of water, being more copious, should saturate the ground for a greater distance, and that, with considerable intervals during which there is no flow, there would be a complete aeration of the ground. It was put in November, 1875. Its effect on the lawn-growth has not been especially marked, but it has thus far acted with the greatest regularity, and is a most satisfactory arrangement.

The accompanying illustration (Figure 16) shows the construction of Field's patent self-acting flush-tank (here referred to), which is intended to be placed immediately outside of the walls of the house and to receive all of its liquid wastes. It is made entirely of earthen-ware or cast iron. The liquids pass through the grating of the pan *B*, and are discharged through a trap that prevents the contained air of the vessel from escaping at the surface. *C* is a ventilating pipe to carry this contained air to the top of the house. *A* is a vessel holding a certain amount of water which has no escape save through the siphon *D*. When the chamber is entirely filled, the pouring in of a few

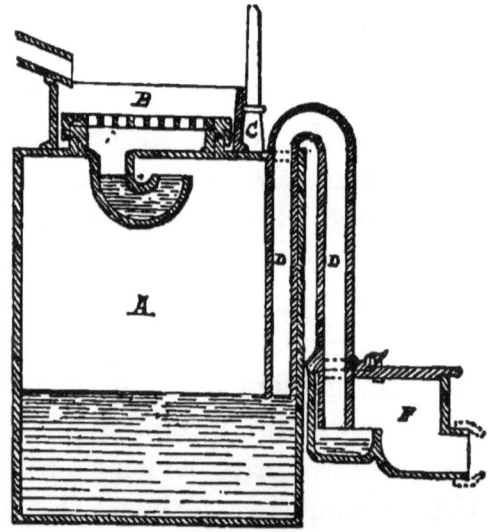

Figure 16.

extra quarts of water, which is sure to occur sometime during the day, brings the siphon into action, and it flows copiously until the chamber is empty to the depth below which solid matters are permitted to accumulate, to be occasionally cleared out on removing the pan *B*.

The purpose of this apparatus is to prevent the constant trickling away of the small stream usually flowing from the house with too little movement to carry forward obstructing matters, such as are sure sooner or later to clog any ordinary house drain. It also furnishes a sufficiently strong flow to secure a wide distribution of the liquid instead of allowing it to soak slowly into a small area of soil. From its intermittent action, also, it fills the ground for a short time, and then as the liquid subsides fresh atmospheric air enters the soil and assists, by its oxidizing action, in the work of purification. Whether the irrigation be on the surface or by means of underground pipes, this copious intermittent discharge is in every way preferable to the steady small flow.

English engineers who have paid much attention to this subject, seem to have settled on this intermittent application of sewage to the soil, with the accessory, in the warmer and more dangerous seasons, of the action of the roots of plants. as the best means for defecating all liquid wastes.

At a recent meeting of the French Horticultural Society, there was a discussion as to the influence of plants on water containing putrefying organic matter; and evidence was adduced to show that while such water left to itself retains its putrescent character, the same water in which the roots of growing plants are feeding loses entirely the *bacteria* which accompany putrescence and contains only the larger infusoriæ which are peculiar to wholesome water. It was sufficient to allow a living root to act for five days for the water to lose all its bad smell and to become purified.

HOUSE VENTILATION.

Incidentally to the seclusion of sewer air from our houses, we have to consider the subject of general ventilation, — a subject that has been more bemuddled and befogged by *quasi* scientific treatment than any other connected with domestic life, unless it be the much vexed and generally misunderstood subject of sewerage itself.

The best practical statement I have met about ventilation was contained in the remark of a mining engineer in Pennsylvania: " Air is like a rope ; you can pull it better than you can push it." All mechanical appliances for pushing air into a room or a house are disappointing. What we need to do is to pull out the vitiated air already in the room ; the fresh supply will take care of itself if means for its admission are provided.

It has been usual to withdraw the air through openings near the ceiling, that is, to carry off the warmer and therefore lighter portions, leaving the colder strata at the bottom of the room. This serves to purify the air but it is very wasteful of heat, and causes too great variations of temperature above and below. Much the better plan would usually be to draw this lower air out from a point near the floor, allowing the upper and warmer portions to descend and take its place.

An open fire with a large chimney throat, is the best ventilator for any room ; the one half or two thirds of the heat carried up the chimney is the price paid for immunity from disease ; and large though this seems from its daily draft on the woodpile or the coal-bin, it is trifling when compared with doctors' bills and with the loss of strength and efficiency that invariably result from living in unventilated apartments.

In ventilation, care should be taken to avoid drafts, not alone from the danger of taking cold as a consequence of sitting in a draft, but even more from the fact that persons inconvenienced by currents of air close the ventilating apertures as the easiest means of getting relief, and so subject themselves to contaminations of the atmosphere which, in addition to their other bad effects, are far more active in producing colds than even the drafts themselves. Dr. De Chaumont in his papers on Habitations says : —

" Usually a current at the rate of one and a half to two feet per second, equal to walking through still air at the rate of one to one and a half miles per hour, is hardly perceived ; two and a half to three feet per second is distinctly perceptible (equal to walking one and three quarters to two miles per hour in still air) ; and four to five feet per second (equal to walking two and three quarters to three and a half miles per hour) is a positive draught. Our object, therefore, ought to be in supplying an air space with fresh air, to take care that the current should nowhere exceed two feet per second in the room itself, and should be kept as near as possible at five feet per second at the point of entrance. I have already shown that 3,000 cubic feet per hour are necessary for each occupant of an air space to preserve good hygienic conditions, and it will therefore be easy to calculate the size of the openings required for entrance. If we had an opening of one square foot of area, and the air coming in at five feet per second, this would obviously give us five cubic feet of air per second, or, as there are 3,600 seconds in an hour, 5×3,600=18,000 cubic feet of air per hour. But as we ask for only 3,000 cubic feet for each person, one sixth of this will be enough, $\frac{18,000}{3000}=6$, so that instead of the opening requiring to have a sectional area of one square foot, it will be necessary to have only one sixth of a square foot, or twenty-four square inches. Again, it being quite obvious that if a certain amount of air finds its way in, an equal bulk of air must find its way out, we must have at least as much opening for the exit as for the entrance of air, so that altogether the sectional area of ventilation openings require to be one third of a square foot, or forty-eight square inches for each occupant of an air space."

The shape and arrangement of ventilating tubes is very important. The aim should always be to reduce friction as much as possible; which is to be accomplished by giving the smallest possible circumference to the area of the air space. Round pipes have less circumference to their sectional area than have oval ones, and square pipes have the same advantage over those of oblong section.

The following table serves to show the relative friction in pipes of different forms.

204 SANITARY DRAINAGE OF HOUSES AND TOWNS.

Shape of Section.	Area.	Total length of periphery.	Coefficient of friction, circle being 1.000.
Circle	100	35.45	1.000
Ellipse, E=½	100	35.67	1.006
Duodecagon	100	38.86	1.012
Heptagon	100	36.72	1.036
Hexagon	100	37.22	1.050
Pentagon	100	38.12	1.075
Square	100	40.00	1.128
Rectangle (sides 4 : 5)	100	45.41	1.231
Equilateral triangle	100	45.59	1.286
Right angled isosceles triangle	100	48.28	1.362

Original Area of Single Opening.	No. of parts into which the original opening is divided.	Area of each part.	Total area of parts.
1 square foot	2	0.707	1.414
1 square foot	3	0.574	1.732
1 square foot	4	0.500	2.000
1 square foot	5	0.459	2.236
1 square foot	6	0.408	2.449
1 square foot	7	0.378	2.646
1 square foot	8	0.353	2.828
1 square foot	9	0.333	3.000
1 square foot	10	0.316	3.162
1 square foot	11	0.302	3.317
1 square foot	12	0.289	3.464

These tables are taken from the section on Ventilation of Dr. De Chaumont's papers on "Hygiene."

The relative ventilating capacity of openings is in proportion to the square roots of their areas.

One opening of one square foot will deliver twice as much air as will four openings of one fourth square foot each. The foregoing table shows the size that must be given to each of a number of openings to make them equally effective with one opening a foot square.

Ventilation is much more effective through a single pipe than through several pipes having an equal aggregate sectional area.

Every bend in a ventilating tube increases the resistance to the current and the resistance is proportionate to the angle of the bend.

The admission of fresh air to supply the place of that which is withdrawn is an imperative necessity, and in tightly built modern houses cracks and crannies for this purpose are wanting. It is not unusual in modern houses supplied with furnaces, especially where there is no public sewerage, to find such an arrangement of closet and kitchen drains as permits the escape of some of their dangerous gases immediately into, or into the vicinity of, the mouth of the cold-air box which supplies the furnace, and the flues which furnish the interior of the house with its heated air.

In a house warmed by a furnace the supply from the registers is usually sufficient to feed the chimney, and if the furnace chamber draws its air from the outer atmosphere, from an unfouled locality, and by all means *not* from a cellar, the only objection lies in the character of ordinary furnace heating. Concerning this it need be said here only that iron heated by hot water is better than iron heated by the direct action of fire, and that, if water-pipes be not used, wrought iron is a much safer material than cast-iron for the transmission of the heat.

In all houses which are connected with cess-pools or public sewers, especial pains should be taken to supply enough fresh air for the fires through some efficient means of communication with the outer atmosphere. Otherwise, there is danger that they will feed themselves from badly trapped communications with the drain.

Sunlight is the handmaiden of ventilation and fresh air. Indeed, ample sunlight and the avoidance of a damp soil may be taken as the very fundamental conditions of healthy living.

In the lying-in hospital in Dublin the mortality of new-born infants during twenty-five years preceding its ventilation was one in six. In the twenty-five years following the supply of pure air by better ventilation, it was one in one hundred and four.

It seems almost incredible that such striking changes should have taken place so recently, but it is to be remembered that it is only about one hundred years since oxygen was discovered, and hardly fifty years since the physiology of respiration was made known; while the fact of injury from breathing foul air is indeed a very recent discovery.

PRIVATE DRAINS.

As with almost every department of sewerage work so with private drains; the engineer in charge endeavors to combine in his rules for distinctive points all that experience has developed. And as the work in Providence is among the most recent in this country, I give herewith the regulations proposed by Mr. Shedd and adopted by the city government. He prefaces them with the following remarks: —

. . . .

"No amount of skill, care, and expense in building the public sewers will relieve the property-holder from the necessity of constructing his private drains with all possible care. These drains often cause, in the aggregate, more trouble, on account of imperfect plan and construction, than all the rest of a sewerage system.

. . . .

"Although house-drains are laid at the expense of the owners of the premises to be drained, the 'Rules' require the work to be done under the permission and supervision of the Water Commissioners, and under the immediate inspection of their engineer of private drains; also that it shall be done by a licensed drain-layer, under bond to do faithful work. These restrictions have been proved by experience, in many cities, to be necessary to secure housekeepers from the great annoyance to which they would otherwise be frequently subjected from imperfect arrangement or unfaithful execution."

"1. Applications for permits to connect with any sewer which has been constructed, or which is in process of construction, by a committee appointed by the Board of Aldermen, must be made in writing to the Water Commissioners by the owners of the property to be drained, or by their duly authorized attorneys, and must be accompanied by a clear description of the premises to be drained, and of the drains required, and also by certain agreements, all as provided in the printed form of application issued by said commissioners.

" 2. No one but a drain-layer, duly licensed by the water commissioners, will be allowed to make connections with the public sewers named in the above section, nor to lay any drains in connection therewith.

" 3. At least twenty-four hours' notice must be given at the office of said commissioners before any street or public way can be opened for the purpose of laying a private drain.

" 4. No drain-pipe can be extended from work previously done and accepted, or new connections of any kind be made with such work, unless previous notice of at least twenty-four hours is given to the engineer in charge of private drains.

" 5. No work of laying drains can be commenced or continued unless the permit is on the ground in the hands of the drain-layer, or some one employed by him.

"RULES FOR LAYING DRAINS.

" 1. In opening any street or public way, all materials for paving or ballasting must be removed with the least possible injury or loss of the same, and, together with the excavated material from the

THE DETAILS OF HOUSE DRAINAGE. 209

trenches, must be placed where they will cause the least practicable inconvenience to the public. As little as possible of the trench must be dug until the junction-piece into the sewer is found, unless it is first determined to make a new opening into the sewer.

"2. Whenever the sides of the trenches will not stand perpendicular, sheeting and braces must be used to prevent caving.

"3. No pipes or other materials for the drains can be used till they have been examined and approved by the chief engineer or one of his assistants, or by a duly-authorized inspector.

"4. The least inclination that can be allowed for water-closet, kitchen and all other drains of not over six inches diameter, liable to receive solid substances, is one half an inch in two feet; and for cellar or other drains, to receive water only, one quarter of an inch in two feet. All drains to be laid at a grade of not over one half an inch in two feet, between the sewers and the sidewalks.

"5. The ends of all pipes not to be immediately connected with water-closets, sinks, down-spouts, or catch-basins, are to be securely guarded against the introduction of sand or earth by brick and cement, or other water-tight and imperishable materials.

"6. All pipes that must be left open to drain cellars, areas, yards, or gardens, must be connected with suitable catch-basins of brick, the bottoms of which must not be less than two and a half feet below the bottom of the outlet pipe, the diameter not less than three feet, and the form and construction of which are to be prescribed by the officers named in the third rule. When meat-packing-houses, slaughter-houses, lard-rendering establishments, hotels, or eating-houses, are connected with the sewers, the dimensions of the catch-basins will be required to be of a large size, according to the circumstances of the case. When the end of the drain-pipe is connected with a temporary wooden catch-basin for draining foundations during the erection of buildings, the drain-layer will be held responsible that no dirt or sand is carried into the drain or sewer from such temporary catch-basin.

"7. No private catch-basin can be built in the public street, but must be placed inside of the line of the lot to be drained, except when the sidewalks are excavated, and used as cellars.

"8. No privy-vaults can be connected with the sewers except through an intervening catch-basin; and the discharge-pipe of the vault must be high enough above its bottom to effectually prevent anything but the liquid contents of the vault from passing into the drain.

"9. The inside of every drain, after it is laid, must be left smooth and perfectly clean throughout its entire length.

"10. In case it shall be necessary to connect a drain-pipe with a public sewer where no junction is left in such sewer, the new connection with such sewer can only be made either by one of the employees of the commissioners, or when an officer named in rule third is present to see the whole of the work done.

"11. Whenever it is necessary to disturb a drain in actual use, it must in no case be obstructed without the special direction of one of the officers named in rule third. No pipe-drain can be laid above the bottom of a wooden drain, whether in actual use or not, unless the pipe is made to rest either on brick or stone, or other suitable support. In no case will drain-pipes be allowed to rest on wood or other perishable material.

"12. The back-filling over drains, after they are laid, must be puddled, and, together with the replacing of ballast and paving, must be done within forty-eight hours after the completion of that part of the drain lying within the public way, and done so as to make them at least as good as they were before they were disturbed, and to the satisfaction of the commissioners and their engineer; and the owner will be held reponsible for any subsequent settlement of the ground. All water and gas pipes must be protected from injury or settling to the satisfaction of the engineer.

"13. Every drain-layer must inclose any opening which he may make in the public streets or ways, with sufficient barriers; and must maintain red lights at the same at night; and must take all other necessary precautions to guard the public effectually against all accidents, from the beginning to the end of the work; and can only lay drains on condition that he shall use every precaution against accidents to persons, horses, vehicles, or property of any kind.

"14. In case a water or gas pipe should come in the way of a drain, the question of passing over or under the water or gas pipe, or of raising or lowering it, must be determined by one of the officers named in rule third. In no case can the drain-layer be allowed to decide the question himself.

"15. No exhaust from steam-engines can be connected with the private or public drains, and no blow-off from steam-boilers can be so connected, without special permission from the commissioners or their engineer.

"16. Such information as the commissioners have with regard to the positions of junctions will be furnished to drain-layers, but at their risk as to the accuracy of the same.

"17. When any change of direction is made in the pipe, either in a horizontal or vertical direction, curves must be used. No pipe can be clipped in any case.

"18. All persons are required to place an effectual trap in the line of drain just before it leaves the premises, and to make an open connection with a down-spout back of the trap; also to make an open connection with the highest part of the soil pipe within the premises, through a large pipe or flue, to a point above the roof of the building.

"19. Every person violating any of the provisions of the foregoing rules shall pay a fine of not less than twenty nor more than fifty dollars."

CHAPTER VII.

THE DRY CONSERVANCY SYSTEM.

THE cases are by no means few in which the easiest solution of the excrement nuisance problem is to be sought through some form of dry conservancy, that is to say, the admixture of either earth, or coal ashes, or other dry household or town refuse, in sufficient quantities for complete absorption so that the degree of moisture in the material itself may be reduced to a point where a healthy decomposition will be carried on, instead of the foul putrefaction to which the production of offensive and dangerous gases is chiefly due.

The statements in this chapter, so far as they are my own, are based upon an amount of experience and observation sufficient to have brought the conviction that the advantages of the dry system are by no means adequately appreciated either by the public at large, or by those having official direction of such matters. As the most of what is known on the subject is the result of actual experiment, and as the investigations upon which the system must largely rest in seeking public favor, have been chiefly made by officers detailed by the health authorities of England, it has seemed best to insert extracts from foreign health reports, sanitary

214 SANITARY DRAINAGE OF HOUSES AND TOWNS.

journals, etc., giving the accounts of these investigations in the language of those who have made them. An attempt to condense these various statements, while it might lead to the avoidance of repetition, would probably detract from the force of the facts and experience stated.

The systems coming under this head are three in number.
1. Moule's Earth-Closet System.
2. The Goux Earth-Tub System.
3. The Ash-closet, which is largely used in certain manufacturing districts in England.

These will be considered separately with such evidence, favorable or unfavorable, as I have been able to obtain.

MOULE'S EARTH-CLOSET SYSTEM.

The use of the earth in closets, under the methods now so well known, is the invention of the Rev. Henry Moule, Vicar of Fordington, England. It has been subjected to a very active public discussion during ten years past, and has had many trials in public and private establishments, and in one or two cases in whole villages. On the whole, its progress has been quite as rapid and secure as could have been prudently hoped for. Its introduction into this country dates back to the year 1868, and although it has not proved a profitable investment for the company who so energetically presented it to public notice, it is constantly and steadily winning its way as being obviously the best available system for certain circumstances; while the encomiums which it has received from those who have experienced its benefits here, have not been less satisfactory than those which have attended its introduction in Great Britain.

Fortunately I am qualified to write on this subject from large experience extending over the whole period of its use in America, but what I say in its favor is to be accepted in the light of the fact that I was among the earliest of its champions, had a pecuniary interest in its success, and have still (aside from the hope of profit which I fear vanished long ago) a very earnest desire to see it meet with the general recognition to which it seems entitled.

For seven years past I have, at my own residence, depended entirely upon some form of earth-closet, and in my present house have had in operation for five years, winter and summer, two closets in constant daily use, — one on each floor of the house, with such success that I would on no account exchange them for the water-closet which is so universally used among my neighbors. The manner in which these closets are arranged will be described under the appropriate heading, and it may be well to refer here to the description already given (page 196) of the way in which all the liquid wastes of the house are disposed of with the help of a Field's flush-tank and irrigating drains.

Several pages of what follows on this subject are taken from an earlier work written in 1869.

Before the earth system can be adopted into general use, the slight care and attention that its success requires must be accepted as an addition to the details of domestic and municipal economy.

The water system with its enormous bills of expense for reservoirs, aqueducts, service-pipes, plumbing work, and sewers, requires constant supervision and care. Whether in private establishments in the country or in large cities, the details of its management require an amount of supervision and of costly labor which, could they have been set forth before the system was anywhere introduced, would have seemed an insuperable objection to its adoption. Now, they are taken as a matter of course, and water-rates and sewer commissioners' taxes are accepted as a necessity of civilized life, and are paid without demur.

The earth system promises to do away with the necessity for most of these charges, and to produce a money result which will more or less repay the others.

At the same time, the perfect carrying out of the earth system of sewage will require a certain amount of care and some expense, which it will be better to consider at the outset. It is not worth while to make a comparison between the requirements of the two rivals, because the more vital considerations, according to which the verdict is to be given, are so weighty that the question of relative cost is comparatively insignificant.

There are two extreme cases to be considered,

THE DRY CONSERVANCY SYSTEM. 217

and the various conditions that fill the gap between them will necessarily resemble one or the other according to their magnitude. In all cases, the *principles* are identical.

1. The earth for use in closets must be dry; not necessarily dried by artificial heat, but made as dry as it can be by exposure to the air and by the exclusion of rain.

2. It must contain enough alumina (clay), or organic matter, or oxide of iron *or* be sufficiently powdery to give it sufficient absorbing power.

3. It must be sifted of its stones and coarser particles.

4. The mechanical arrangement of the closet must be such that a sufficient quantity of earth will be, with certainty, deposited upon the fæces — enough to cover them, and to absorb the urine of the single evacuation. And the accumulation under the seat must be occasionally raked down or leveled off in the vault when an ordinary vault is used.

5. When the vault or receptacle has become too full, its contents must be removed, and before the supply is exhausted the reservoir must be refilled.

6. If the earth is to be again used, its organic matter must be destroyed by fermentation, and its moisture must be evaporated.

7. In towns, some system must be adopted for the supply of earth and removal of deposits — either by the public authorities or by private enterprise.

1. As Mr. Moule very tersely states the case,

"An earth-closet will no more work without *dry earth* than a water-closet will work without water;" but the dryness here referred to is not absolute dryness, for the earth of the closet will always contain what moisture may be absorbed from the atmosphere. This, and even a little more than this, I have found to be not at all objectionable. What is required is, according to Professor Joy, that so much of the moisture of the fæces shall be immediately withdrawn from them that there shall be too little left to cause an offensive putrefaction.

The best manner for drying the earth depends very much upon the quantity required, and the means at command. Two or three cart-loads, which will be sufficient for a year's use of an ordinary family, may be taken from a ploughed field or a road-side gutter during the dry weather of summer. Dumped in an out-of-the-way corner under a wood-shed, or in any other dry place, being underlaid with boards to prevent it from absorbing the moisture of the earth, it will soon become sufficiently dry for use, and will remain so throughout the dampest and foggiest weather of the winter or spring. It might be equally well kept in a dry and well-ventilated cellar. It may be sifted, little by little, as wanted, and it will answer tolerably well if it is merely put through the ordinary coal-sifter, though something finer would be preferable. My sieve has six meshes to the inch; perhaps four would do as well. When the earth is sifted, it may be stowed away in boxes or barrels in some easily accessible place, and there remain until wanted for use. This is sufficient for the requirements of a private house.

In preparing for the supply of a large town, it would be necessary to procure a land-right, in order that deep excavations can be made. The amount of earth needed will be very large, and it must, of course, be procured in the cheapest way. This will be in nearly all cases, by making a clean sweep as deep as it is economical to work, and making an acre of land produce as much earth as possible. The high price of land in the immediate vicinity of the town may make it desirable to go to a considerable distance, in order to secure cheap land and cheap transportation combined. The earth being procured, the first drying can be most economically done near the spot from which it was taken, by simply storing it under rain-tight and well-ventilated sheds. It would, perhaps, be well to make some provision for rapid, artificial drying in the town to provide against emergency and accident.

2. There is undoubtedly a good deal of difference in the effectiveness of earths of various composition; though, with a considerable range of experiment and observation, I am inclined to think that the kinds of earth which are *not* suited for use in the closet are much fewer than would be generally supposed. Pure sand and gravel are nearly worthless, but I think that any earth that contains enough clay or organic matter for the production of ordinary crops will answer the purpose. A nearly

pure clay, however, is objectionable from its tendency to absorb moisture from the air. If to be used only once, an equal weight of muck or peat may be, from its greater bulk, more valuable than clay. Suitable clay could probably be re-used many more times, and so would be cheaper for use in towns. Without being able to give a definite scientific reason for the opinion, I think that a clay loam, highly charged with oxides of iron (notably reddish clay loams), would be the best. In my own experience, I have found anthracite coal ashes to answer a perfectly good purpose, — especially after one use in the closet has dampened them enough to lay their dust.

3. The sifting of the earth is, as I have shown, a very simple matter when it is a question merely of the supply of a single household. When large quantities are required, it would be the most economical plan to adopt revolving screens, such as are used for cleaning coal at mines, the construction being similar to that of the bolting screen of a common flour-mill. Such a screen should be, probably, twenty feet long, the first half of its length being furnished with quarter-inch meshes, and the next with half-inch meshes. Stones and very large lumps would be discharged at the end of the screen; the coarser particles passing through the half-inch mesh might be broken up in a stamping-mill and resifted. If the screening-house were built in a side hill, so that carts could lead directly to the screen, and the prepared earth fall to a story

below, much necessity for shoveling would be obviated.

4. Concerning the mechanical arrangement for the closet, I am more and more inclined to the opinion that Mr. Moule's device is the only one that will be effective under all circumstances. Possibly variations in the size of the "chucker" (by which the quantity of earth used is measured), according to the quality of the earth, may be found to be desirable. Whether this apparatus is used or whether we depend on covering with a hand-scoop, the quantity should be regulated by the quantity of urine to be absorbed, and at each urination earth should be thrown down, to prevent undue moisture.

In an ordinary broad vault the deposits will naturally form a heap under the seat. This must be, now and then, leveled off, and the surface exposed by the leveling should be thinly covered with the drier earth near the sides of the vault. Probably under no ordinary circumstances would it be necessary to do this oftener than twice in a month. In the commode and the up-stairs closet, it will never be necessary. With the Broadmoor tank, or larger vault, it will be.

5. Just as it is requisite to empty a cess-pool, or fill the tank over a water-closet, as occasion requires, so it is necessary to supply fresh earth to the earth-closet, and carry away the accumulation. The details of this work are too simple to need attention here.

In the case of towns, where the system is in anything like general use, the care of the closets should devolve almost exclusively upon associations or individuals engaged in the business of earth supply. Having, as a gardener, undertaken this in Newport for the sake of the manure to be obtained, I am already convinced that in all places where manure has even a moderate value, it will be unnecessary to make a charge for the earth and attendance. The preparation of the earth and the amount of transportation constitute a trifling tax when compared with the value of the product. When the business increases, so that the time of a man and a horse and cart will be constantly employed, the details can be somewhat simplified, and the rounds made with more regularity; the only precaution necessary being, to have always an abundant supply of earth ready in advance, so that protracted wet weather will not require regular delivery to be postponed in order to make use of the first fair weather for securing earth.

Wherever the demand is sufficient for the business to be regularly systematized, the earth may be delivered as ordered, just as coal is now delivered from coal-yards, and it would be proper to make a charge for "carrying in," as in handling coal. If the cart is suitably covered against rain, it is most convenient to carry the earth in bags. These may be emptied into a bin in the cellar, from which commode-hods are supplied, or into the hoist-box of the up-stairs closets, or they may be carried to closets on the upper floors of houses.

The deposits may be removed in baskets, and emptied into the cart on its returning rounds. Barrels are too heavy for one man to handle, and are less convenient than bags for filling closet-reservoirs.

In places where manure has not sufficient value to pay the cost of attendance, the charge necessary to make a profitable business of attending to a considerable number of closets would be much less than the water rates and plumbers' bills that are an inseparable part of the water system. If ashes are used, the addition of the closet manure to them will not materially increase the cost of their handling, and it will give them a value which they do not now possess.

6. In the country where the manure is to be applied directly to the garden, it will be perhaps better to use the earth but once, as there is an advantage in having it as bulky as possible for more even distribution; but even in this case it should not be applied in its fresh state. It should be first thrown into a bin or into barrels, in which it will retain its moisture long enough for perfect fermentation. In this way its paper will be destroyed, and its fæcal matter will be diffused throughout the mass and absorbed by the earth; while the earth itself will have its own fertilizing constituents developed by the decomposition going on within it. When ready for use, the earth will be nearly indistinguishable from that freshly taken from the field; but its manurial power will be materially increased. If the

manure is to be sold in the market or is to be transported to any distance, it should be repeatedly used, in order that its value may be as much as possible increased. The deposits taken from the closets should be carried to the earth depot, thrown into compact heaps, moistened a little, if necessary, and left to ferment. After a sufficient time, these heaps may be shoveled over, and left to undergo a second fermentation. They may then be spread out to dry, or, better, removed to a drying-room where there is a free circulation of air. After becoming dry, the earth may be passed through a screen, and the finer parts stored away for further use; the small amount of coarser matter may be again moistened and fermented. Of this latter, the quantity will be very small, and it will consist chiefly of dried-up solid fæces, which it may be found best to pulverize and use directly as manure, or it may be mixed with deposits freshly brought in from the closets. It will help the fermentation of these, and will be entirely absorbed.

7. What is the best arrangement for towns and villages it is now too early to say; but in any case the details of the system would be simple and easy of execution. If the value of the manure is enough to make the earth business a source of profit, it may be safely left to private enterprise; but even in this case the sanitary authorities of the town should provide for the inspection of closets, especially among the poorer classes, and it should be required that all comply with such provisions as the public interest makes necessary.

If the preservation of the manure is not an object, the removal of the accumulations may be provided for, as is now done in the case of ashes, etc. The public authorities should, in all cases, assume such control of the matter as to insure the perfect working of the system; but the manner in which private establishments shall be supplied with earth is a question to be decided by the peculiar circumstances of each case. Just as no water-closet should be allowed to remain in use without a supply of water or with an obstructed soil pipe, so should no earth-closet be allowed to become ineffective from the neglect of its owner to provide it with earth or to have its accumulations removed. It is now necessary, in even the smallest towns, to prevent any outrageous neglect of common privies; and the extension of the same system of inspection to meet the requirements of the dry-earth sewage would be neither difficult for the authorities nor onerous to householders.

THE MANURE QUESTION AS AFFECTING THE EARTH SYSTEM.

In this connection, the following, which I published in 1872,[1] seems worth reproducing: —

It is a very difficult matter to fix the value of any animal manure, except by accurate analysis of each separate sample. Opinions as to the value of human excrement vary widely according to the standard of comparison taken. It is a singular fact (which does

[1] *Earth-Closets and Earth Sewage*, p. 49.

not obtain with reference to most other manures) that the valuation of human excrement made by chemists is very much less than that of the practical farmer. For instance, in England, where the value of the mineral constituents of the material seems to be entirely disregarded, it is usual to measure the value solely by the amount of *ammonia* that may be produced from it; those parts of night-soil which are the key to the lasting fertility of the land are not taken into the account, and ammonia alone (which, although a most valuable and efficient aid to the farmer, counts as nothing in giving *permanent* fertility) is considered.

. . . .

In this connection, I reproduce a portion of an article of my own written for Judd's " Agricultural Annual " of 1868 : [1] —

" The average population of New York city — including its temporary visitors — is probably not less than 1,000,000. This population consumes food equivalent to at least 30,000,000 bushels of corn in a year. Excepting the small proportion that is stored up in the bodies of the growing young, which is fully offset by that contained in the bodies of the dead, the constituents of the food are returned to the air by the lungs and skin, or are voided as excrement. That which goes to the air was originally taken from the air by vegetation, and will be so taken again. Here is no waste. The excrement contains all that was furnished by the mineral ele-

[1] " Sewers and Earth-Closets, and their Relation to Agriculture."

ments of the soil on which the food was produced. This all passes into the sewers, and is washed into the sea. Its loss, to the present generation, is complete.

"In the present half-developed condition of the world, there is no help for this. The first duty in all towns is to remove from the vicinity of habitations all matters which, by their decomposition, would tend to produce disease. The question of health is, of course, of the first importance, and that of economy must follow it; but it should follow closely, and perfect civilization must await its solution.

"Thirty million bushels of corn contain, among other minerals, nearly seven thousand tons of phosphoric acid, and this amount is annually lost in the wasted night-soil of New York city.[1]

"Practically, the human excrement of the whole country is nearly all so disposed of as to be lost to the soil. The present population of the United States is not far from 35,000,000. On the basis of the above calculation, their annual food contains over 200,000 tons of phosphoric acid, being about the amount contained in 900,000 tons of bones, which, at the price of the best flour of bone (for

[1] Other mineral constituents of food — important ones, too — are washed away in even greater quantities through the same channels; but this element is the best for illustration, because its effect in manure is the most striking, even so small a dressing as twenty pounds per acre producing a marked effect on all cereal crops. Ammonia, too, which is so important that it is usual in England to estimate the value of manure in exact proportion to its supply of this element, is largely yielded by human excrement.

manure), would be worth over $50,000,000. It would be a moderate estimate to say that the other constituents of food found in night-soil are of at least equal value with the other constituents of the bone, and to assume $50,000,000 as the money value of the wasted night-soil of the United States.

"In another view, the importance of this waste cannot be estimated in money. Money values apply, rather, to the products of labor, and to the exchange of these products. The waste of fertilizing matters reaches farther than the destruction or exchange of products — it lessens the ability to produce.

"If mill-streams were failing year by year, and steam were yearly losing force, and the ability of men to labor were yearly growing less, the doom of our prosperity would not be more plainly written than if the slow but certain impoverishment of our soil were sure to continue.

"Fortunately, it will not continue always. So long as there are virgin soils, this side of the Pacific, which our people can ravage at will, thoughtless earth-robbers will move West and 'till' them. But the good time is coming, when (as now in China and Japan) men must accept the fact that the soil is not a warehouse to be plundered — only a factory to be worked. Then they will save their raw material, instead of wasting it; and, aided by nature's wonderful loom, will weave, over and over again, the fabric by which we live and prosper. Men will build up as fast as men destroy, old matters will be reproduced in new forms, and, as the decaying forests feed the growing wood, so will all consumed food yield food again.

"The stupendous sewers which have just been completed in London at a cost of $20,000,000, and which challenge admiration as monuments of engineering achievement, are a great blessing to that filth-accursed town, and, in the absence of anything better, they might, with advantage, be imitated elsewhere. They have had an excellent effect on the health of the population, by removing a prolific cause of typhoid fever and other fatal diseases. As affording needed relief from malaria, they are of immense importance. Still, they are a great (although necessary) evil, inasmuch as they wash into the sea the manurial products of 3,000,000 people, to supply whom with food requires the importation of immense quantities of grain and manure.

"The wheat market of one half the world is regulated by the demand in England. She draws food from the Black Sea and from California; she uses most of the guano of the Pacific islands; she even ransacks the battle-fields of Europe for human bones, from which to make fresh bones for her people; and, in spite of all this, her food is scarce and high, and bread-riots break out in her towns.

"An earnest effort is now being made to use the matters discharged through these sewers for the fertilizing of the lands toward the eastern coast. For this purpose, it is intended to build a sewer, forty miles long and nine and a half feet in diameter, which, with the incidental expenses of its construction and management, will cost about $10,000,000. The sewage company have a farm at Barking, on which they have experimented very successfully, one acre of their irrigated meadows having produced nine tons of Italian rye grass in twenty-two days, and fifty tons during the past season up to August 15, with a prospect that the yield for the whole season will be, at least, seventy tons from a single acre.

"The system of sewage irrigation has earnest adherents, and equally earnest opposers. It does seem a pity that, for every pound of excrement that is given to the land, three or four hundred pounds of water must go with it; and it is probable that such highly diluted manure can be used with advantage only on grass crops. It is further asserted that, as the best results can be obtained only by the application of from 6,000 to 10,000 tons of the liquid per acre, the cost of the process must prevent its general adoption. However, the scheme is about to be thoroughly tested, and it is to be hoped that its success will be such as to secure a return to the soil of a vast amount of valuable matter which, hitherto, has been worse than thrown away.

"The many attempts that have been made to extract the fertilizing parts of the sewage from the deluge of water with which they are diluted, have entirely failed of their object. If, as now seems probable, the best and cheapest way to remove waste matters from large towns is by dilution in large quantities of water, the efforts of agriculturists must be directed to the best means of making use of the mixture."

. . . .

Wherever I have used either the earth from the closet or the contents of my filtering-casks, the effect has been obviously much greater than it would have been from the use of the raw material alone. A portion of the improvement, no doubt, is due to the more even distribution that the increased bulk makes possible; but I am inclined to attach much greater importance to a suggestion contained in an article prepared by Colonel Weld for the " Agricultural Annual" for 1870. He says:—

" Most soils contain a much larger quantity of substances required by the plant than would be available in several years' cropping. These are gradually rendered soluble and fit for plant-food by weathering year by year. The result of mingling a soil with manure which is undergoing active fermentation is to cause decomposition to go on in it more rapidly, and so it is certain that a part of the benefit arising from the use of soil as an absorbent in stables is that a larger supply of plant-food is prepared from the soil and distributed with the manure."

Of course this introduces an element of uncertainty into the calculation, as it is not likely that any two soils would yield exactly the same fertilizing value to the action of decomposing manure; but it is undoubtedly true that any earth not positively barren will be very beneficially affected by the active decomposition of fæces and urine within its mass.

The same article contains the following lucid statement of the effect of earth on decomposing organic matter: —

"The earth-closet depends for its working upon the deodorizing and absorbing qualities of dry earth. The earth absorbs moisture because it is dry; it absorbs odors, both on account of its chemical nature and the mechanical arrangement of its particles. Earth is sometimes considered as antiseptic, because it so thoroughly destroys some of the products of decay, especially evil odors; but it really promotes decay very energetically. If we lift a piece of cloth, a part of which has been buried by accident for a few days or weeks, we find the part under the earth greatly injured, or entirely rotten, from the contact of earth and moisture. Very dry earth is somewhat antiseptic on account of its dryness.

"The disorganization or decay which earth promotes does not affect living organisms of either plants or animals.[1] Hence, seeds, roots, bulbs, insect life, and the eggs of many birds, reptiles, and insects, are preserved, if buried in earth of natural dryness, so long as life remains.

"The purifying and deodorizing properties of the soil are familiar to almost every farmer boy in the

---

[1] This does not apply, either necessarily or very probably, to the living germs of infection (if such germs there be), as these seem to multiply only under the influence of putrid decomposition.

country. A very slight burying prevents the odor of a decaying carcass being noticeable; a thin covering of earth suffices to suppress the odors of a fresh manure heap; and the most disgusting of all common smells, that of the skunk, may be entirely removed from articles of clothing, or other things contaminated by it, by burying them in the ground a few weeks (of course, absolute contact of the earth with the garments should be prevented, or they would be rendered useless through decay)."

There is one consideration connected with this branch of the subject which is of even greater importance than the mere money value of the single application of the manure. Our present system is one of constant waste. We draw from the soil a certain amount of plant-food with every crop that we grow. In so far as the crop is consumed by man, this plant-food is practically wasted. In the next crop that the land produces, fresh elements are required, and these, in their turn, are thrown away. And so we go on, year after year, always drawing out more than we put back, to the extent of almost the entire food of our population. Of course much of this material finds its way, sooner or later, to the soil; for even that which is washed into the sea may be reclaimed in sea-weed used as manure, or in fish that is used for food. But as a broad proposition, it may be assumed that practically the food of our population returns almost nothing to the soil.

The relief that the earth-closet offers in an agricultural point of view is not to be measured by the simple fact that it furnishes nutriment to the crop to which the manure is applied; for still greater importance attaches to the permanent benefit to the soil, resulting from any system, by which all that has been contributed to plants is surely returned. Instead of removing mineral plant-food with every crop, and sending it to the four corners of the earth, these mineral matters are returned in a form suitable for immediate use. The crop is fed, not by new contributions from the soil, but by the very material which has fed previous crops. The same elements may be used over and over again *ad infinitum*.[1] At the same time, the action of the weather upon the soil, the action of the feeding-roots of plants on the surfaces of its particles, and the power of organic matters (both the added manure and the decaying roots of previous crops) to develop latent fertilizing elements of the earth, all tend to add, year by year, to the active fertility of the land.

DR. VOELCKER ON THE VALUE OF EARTH-CLOSET MANURE.

In 1872, Dr. Voelcker, chemist to the Royal Agricultural Society of England, contributed to its Journal (2d Series, vol. viii.) a carefully considered essay on this subject. The following quotations will show the very favorable light in which he regards the system generally, and the small value he attaches to its product.

[1] Of course, it may or may not be the identical elements it has previously yielded which are returned to a given field, and it is not material to the argument whether it be these or their equivalent.

"In the country — in small county towns, and in isolated establishments, such as county prisons, workhouses, and asylums — the disposal of human excretal matters presents no great difficulty; but their removal from towns is generally attended with considerable expense, no matter what particular system the authorities may adopt. By degrees the town authorities are learning the disagreeable lesson that materials which are excellent fertilizers when safely incorporated with the soil are a nuisance in a town, and cause expenses that are all the greater the more completely the plan of removal accords with the requirements of civilization. . . . . Nothing effects so complete and rapid a deodorization and disinfection of putrid animal matter of every kind, as a well aerated soil.

"Bousingault has shown that there is a larger proportion of oxygen in the air condensed between the particles of a porous soil than in the atmosphere above the land. In the condensed condition in which oxygen exists in a porous soil, it no doubt acts much more powerfully in oxidizing organic matters than the free oxygen of the air. . . . .

"There is no oxidizing agent equal to a porous soil, which is always at hand in almost unlimited quantities, and equally effective in destroying animal effluvia, and the permanently prejudicial properties of excrementitious matter of every description. Few axioms are so true as that which enforces the propriety of returning to the land the fertilizing materials which are removed from it in the produce. In other words the nuisance of a town population ought to be utilized on the land for the production of food.

. . . .

" Mr. Moule has the merit of having given to the public a simple and ingeniously contrived apparatus, which is capable of doing good service in many places; more especially in sick rooms, public establishments such as county prisons and unions, and country houses, where a good supply of water cannot be commanded; and credit is due to him for having pointed out the *repeated* action, and consequently the fitness for repeated use of the same earth.

. . . .

" Where the earth required for absorption can be readily procured in a dried and sifted state, and the land for the utilization of the compost is in close proximity, the earth-closet system recommends itself as a thoroughly efficient plan of disposing of human excreta and the utilization of their fertilizing constituents at the smallest expense and in some cases even with economy."

Composition of a sample of earth-closet manure, used four times in succession, and dried: —

Moisture (loss on drying at 212° Fahr.)	1.49
*Organic matter and water of combination	6.56
Oxide of iron and alumina	14.57
Tribasic phosphate of lime (bone-phosphate)	1.46
Carbonate of lime	9.47
Magnesia	2.20
Potash	1.31
Chloride of sodium	.82
Insoluble silicious matter (clay)	62.12
	100.00
*Containing nitrogen	.39
Equal to ammonia	.47

THE DRY CONSERVANCY SYSTEM. 237

As an illustration of the practical working of the system on a large scale, he gives the following extract of a letter from Captain Armytage, the governor of the West Riding Prison, Wakefield, where as many as 776 earth-closets were in use, the system having been first introduced in the summer of 1866.

"We use the ordinary Moule's closet, or a still simpler box, where the earth is applied out of a small scoop by hand, instead of the self-acting machinery of Moule's closets, which, with ordinary care, acts very well. You must be aware what class of men and women we have to deal with in working out experiments; and I can only say that, after more than three years, I am satisfied that the dry-earth plan is the only sound system that can be worked out, especially among the lower classes and in towns, my principle being to keep all sediments out of the drains. The urine now is collected into tanks, and is sold, or used for manuring the ground, or is thrown upon the earth compost. We find an absence of all smells, that formerly were quite overpowering; and even in the manipulating shed no smell can be discerned, except at the time of turning the compost, and then the smell perceptible in the shed is more that of a Peruvian guano-shed than anything else."

It appears that: "In the course of the year from fifty to sixty tons of earth-manure are obtained, which is chiefly used on the prison grounds. In 1870, about twelve tons were sold at £1 per ton, when the earth was once used; £2 when twice used, and £3 thrice used. It has had remarkable success in growing onions, and has been used with advantage for potatoes, vegetables, and garden produce in general. Half a ton per acre of the earth used once in the closets has also been successfully applied to grass land, and one ton per acre produced two tons, three cwt. of hay. In a second experiment, one ton of the earth manure (once used), produced two tons, two cwt. of hay."

In Voelcker's further chemical investigation of this subject there is given the following analysis made by Dr. Gilbert: —

	EARTH.		
	Before use.	Using once.	Using twice.
Percentage of moisture in air dried and sifted soil (loss at 210° Fah.)	8.440	9.970	7.710
Percentage of nitrogen in air dried and sifted soil	.067	.216	.353
Percentage of nitrogen in soil dried at 212° Fahr.	.073	.240	.383

On the strength of which, it is stated that: "earth passed three times through the closet, in a perfectly dry state, was worth only 6s. 2½d. more per ton than dry garden mold of the composition of the soil employed in the experiments."

In support of his opinion that the value of human excrements has always been popularly overestimated, Dr. Voelcker adduces the following : —

" In 1864, the Prussian government commissioned Messrs. C. v. Salviati, O. Röder, and Dr. Eichhorn, to investigate the modes of collection, removal, and utilization, in various continental towns; and in their report, the Prussian commissioners, who visited various towns in Belgium, France and Germany, showed not only that the householders seldom realized anything like a franc per head per annum for their excretal matter, but that, in the majority of towns, they had to pay something for the removal. It is surprising that in the face of the reports of individuals who have investigated the subject on the spot, and in spite of reliable official reports, embodying the results of personal observations, and dealing with plain matters of fact, many people should still give credence to the unwarranted statement that in Belgium excretal matters are sold at £1 per head per annum, and that most continental towns derive a more or less considerable income from the sale and utilization of human excreta. In the endeavor to correct the erroneous and exaggerated notions which not a few persons entertain with regard to the money value of human excrements, I have purposely confined myself to a statement of facts, which every one may verify who will take the trouble to visit continental towns and make inquiry into the manner in which human excreta are disposed of, and what is realized by the towns by their utilization. The practical conclusion to which an unbiassed inquirer into this subject will arrive is that, as far as the inhabitants of towns are concerned, human excreta are a nuisance, for the removal of which, in most towns they have to pay something.

· · · · · · · · ·

240 SANITARY DRAINAGE OF HOUSES AND TOWNS.

"Speaking generally, solid human excreta, as they leave the body, contain one fourth of dry matter and three fourths of water. The dry matter contains about 1½ per cent. of nitrogen, and 1 per cent. of phosphoric acid.

. . . .

"If it were possible to dry fæces without loss in fertilizing matters, and without the addition of bulky material, they would, in a dry state, be a very valuable manure, for in that state they would contain : —

*Organic matter	88.52
Insoluble silicious matter	1.48
Oxide of iron	.54
Lime	1.72
Magnesia	1.55
Phosphoric acid	4.27
Sulphuric acid	.24
Potash	1.19
Soda	.31
Chloride of sodium	.18
	100.00
*Containing nitrogen	6.00
Equal to ammonia	7.28

"It appears from the preceding figures that, in a perfectly dry condition, two tons of solid human excreta are worth almost as much as one ton of Peruvian guano; and it seems a great pity that a manure possessing such a fertilizing value should be wasted as at present it is in a great measure.

"Still more valuable as a manure is human urine, for its principal constituent — urea — contains nearly fifty per cent. of nitrogen; and uric acid — an active constituent of urine — contains about thirty-three per cent. of nitrogen ; and besides these nitrogenous organic matters, human urine contains a good deal of phosphoric acid. . . . .

" Fresh urine contains, on an average only three per cent. of solid matter, and, according to Professor Way's analysis just quoted, consists of : —

Water	97.000
*Organic matter	2.026
Insoluble silicious matter	.003
Oxide of iron	.002
Lime	.018
Magnesia	.014
Phosphoric acid	.040
Sulphuric acid	.014
Potash	.055
Chloride of potassium	.162
Chloride of sodium	.566
	100.000
*Containing nitrogen	.58
Equal to ammonia	.71

" Notwithstanding this large proportion of water, *the amount of solid matter in the urine voided in a day is just about one third greater than the amount of dry matter in the daily solid evacuations.* It is not easy to calculate with great precision what is the total amount of fæces and urine which is produced by a mixed population of adults and children of both sexes; but it may be safely stated that the amount of dry matter in the solid and liquid excreta of a mixed population does not exceed fifty-six pounds per head, per annum, and that probably it is not more than forty-five or forty-six pounds. . . . .

" On calculating the amount of ammonia which

will be produced on the decomposition of the dry matter of the solid and liquid excrements of each person per annum, we obtain from the

	Ammonia·
23 lbs. of dry matter contained in the solid excreta	1.60 lbs.
And in 34 lbs. of dry matter contained in the liquid excreta	8.12 lbs.
Or altogether	9.72 lbs.

"In other words, five sixths of the ammonia capable of being generated on the decomposition of human excreta is furnished by the urine. By a similar calculation I find that, according to the preceding data, each individual would furnish about 5½ lbs. of phosphates per annum. For simplicity's sake we may assume that each person of a population produces in the solid and liquid excreta 56 lbs. of dry matter per annum. These 56 lbs. produce in round numbers 10 lbs. of ammonia, and 5½ lbs. of phosphates.

"In order to avoid the appearance of a wish to undervalue the intrinsic fertilizing value of human excreta, I would allow $9d.$ per lb. for ammonia, and $2d.$ per lb. for phosphates, and further $9d.$ for the money-value of the remaining constituents, which is rather more than the latter are really worth.

"The excreta of each person of a population accordingly would be worth per annum $9s.$ ($2.16 gold), allowing for —

10 lbs. ammonia at $9d.$ per lb.		$7s.\ 6d.$
5½ lbs. of phosphates at $2d.$ per lb. . . .		0 11
Other matters		0 7
Total value of human excreta per head per annum . .		9 0

"Bearing in mind that five sixths of the total amount of ammonia in the solid and liquid excreta of man are furnished by the urine, and only one sixth by the fæces, and how small is the proportion of the total urine that is passed at the same time, and that our domestic habits prevent the collection and absorption of the whole of the urine, the intrinsic value of the fertilizing matters which can be practically recovered in Moule's earth-closet, is probably not more than one third of their value, or amounts to only 3s. for each person per annum. In order to recover these three shilling's worth of manuring matters, a large quantity of earth has to be used in Moule's closet.

. . . .

"Assuming that the total excreta of a man can be absorbed by the earth without loss, and that they possess an average value of 9s. per annum, each ton of earth used five times in the closet will be worth 22s. 6d.; but as, practically, about two thirds of the fertilizing matters will be wasted in the urine, which cannot be recovered and absorbed by earth, the value of a ton of earth-closet manure used five times will only be about 7s. 6d."

This paper has such scientific importance and is justly entitled to so much respect on the part of all who know Dr. Voelcker's most efficient services in the cause of improved agriculture, that no full statement of the system can be made without full and faithful reference to his essay.

The correctness of his estimate it would be danger

ous to doubt. If accepted as final it simply states, more in detail, the general proposition that with the earth system, as with all others, the incentive must be a sanitary rather than a commercial one, and those who still hold to the great agricultural value of earth-closet manure, however much discouragement they may receive at the hands of chemists, will have to fall back upon the many instances of results from its practical use, which seem to be far beyond what its theoretical composition would account for.

This paper suggests the proposition in agricultural chemistry that, just as the nitrogenous products of organic decay are consumed by the action of aerated charcoal, so are they consumed by a similar action of aerated earth (or other dry and porous material) and that — in this regard — there may be a loss of organic elements of fertility in the process of " summer fallowing," — a loss that is compensated for by a certain gain, but a loss nevertheless.

MOULE'S APPARATUS.

The original apparatus devised by Mr. Moule is manufactured by a successful company in London, who have, as experience has suggested modifications, added a variety of forms and several improvements, which have greatly extended the scope of the system.

Precisely what the earth-closet and its accessories, as now contrived, accomplish, is the following:

A comfortable closet on any floor of the house,

supplied with earth, and cleansed of its deposits without the intervention or knowledge of any member of the household;

Figure 17.—The Commode. This is a "self-contained" closet.

A portable commode in any dressing-room, bedroom, or closet, the care of which is no more disagreeable than is that of an ordinary fireplace;

Appliances for the use of immovable invalids

246 SANITARY DRAINAGE OF HOUSES AND TOWNS.

which entirely remove the distressing accompaniments of their care; and

The complete and effectual removal of all the liquid wastes of sleeping-rooms and kitchen; and

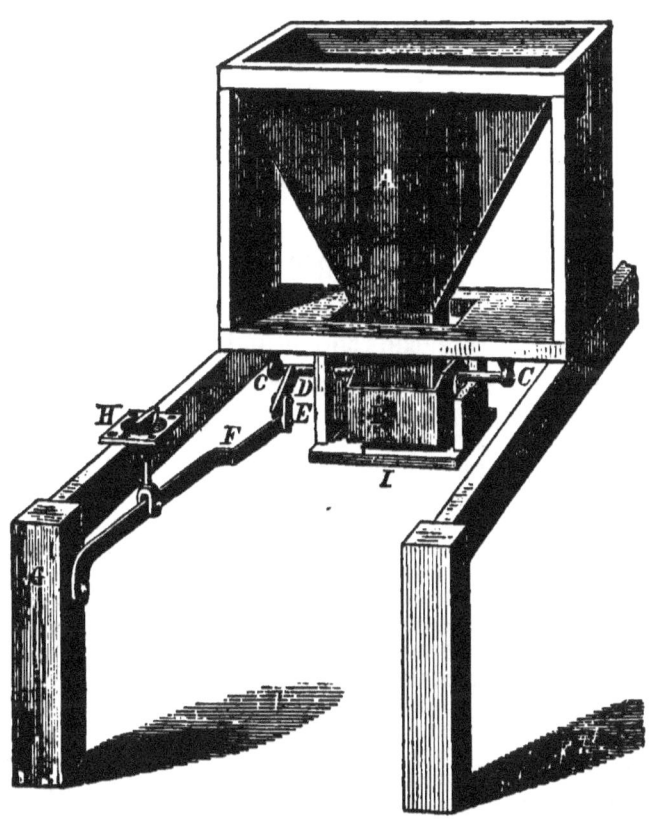

Figure 18.

The complete suppression of the odors which, despite the comfort and elegance of modern living, still hang about our cess-pools and privy-vaults, and attend the removal of their contents.

The simplest form of Moule's apparatus is the portable commode shown in Figure 17, which is adapted for use in any room or closet, and which with only ordinary care is as inoffensive in odor and as convenient and little disagreeable in attention as a common anthracite stove.

The arrangement of the mechanism of this commode is shown in Figure 18.

A, is a swinging hopper, capable of holding an ordinary coal-hod full of earth; B is the "chucker" which on being tilted by lifting the handle H, throws forward the proper quantity of earth into the moveable hod standing under the seat. When the handle is released the chucker drops back into the position shown in the cut, and is filled from the hopper which enters its top, its mouth being at the same time closed by the shelf I, suspended beneath it. The commode should be supplied with two hods, the one not in use being exposed to the air during the time that it is waiting. When fresh earth is needed for the hopper it is carried to it in this dry hod, which after being emptied, is substituted for the filled one under the seat.

An apparatus for more regular use, — for a fixed closet where the circumstances allow of the construction, — is shown in Figure 19.

There is a considerable reservoir for earth built up above the level of the top of the hopper, — my own holds from four to six weeks' supply. In lieu of a moveable hod, there is fixed beneath the seat, reaching through an opening in the floor, a galvan-

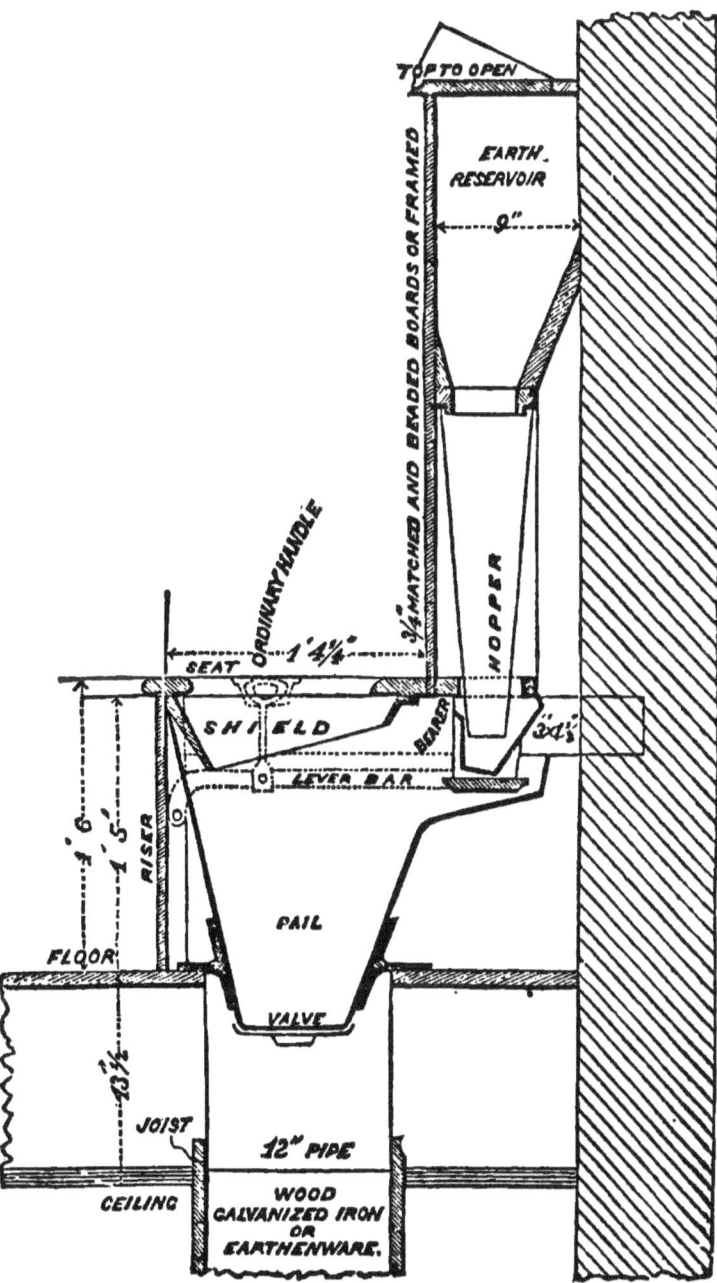

Figure 19. — Apparatus with valved funnel.

ized iron funnel closed at the bottom with a moveable valve. This funnel-mouth opens into a twelve inch galvanized iron pipe which passes to the receptacle in a cellar or room below. In my own case the closet on the main floor of the house delivers through a short pipe into the top of a tight brick vault three feet square and seven feet high. This vault is closed over with brick work at the top, being plastered close around the pipe so that there shall be no opportunity for an escape of its air into the cellar or into the closet above. In a new construction, I should, as a further precaution, provide some means for ventilating this vault, but we have experienced no inconvenience from it in its present condition, and as there is always an abundant supply of earth within it I can imagine no danger from it.

The closet on the second floor is arranged in precisely the same manner, save that the connecting pipe is long enough to reach through from this floor to the top of a vault in the cellar, — about twelve feet. Each of these cellar vaults has a small manhole in the brick work (with timber header and sill), which is loosely bricked up and then well coated with mortar on the outside. The vaults have to be emptied about three times in the year, when these loose bricks are knocked out to give access, — the opening being plastered up again as soon as the work is done.

The only objection that has at any time been found to this arrangement was due to the fact that owing to its conical form, the funnel, when full, does not empty itself without help. This objection is entirely obviated by making it the duty of one of the users to pull the dumping handle at least once every day, — the smaller accumulation discharging itself freely. There is the further advantage in this that the deposits are thus always at a depth to be most perfectly reached and covered by the falling earth.

In my own case, no earth has been brought to the house for three years. During this time we have used over and over again the same material, which consists almost entirely of anthracite ashes. When the vaults are emptied, their contents are simply heaped up in the cellar, where they become sufficiently dry, after a month or so, to be used again. Much of this accumulation has passed through the closet, at least ten or twelve times. Under no circumstances has there been any indication that it is anything but ashes, with a slight admixture of garden soil. In my case, the earth-closet system is entirely free from any complication.

A simple form of earth-privy is shown in Figure 20.

Figure 21 shows an arrangement by which closets on different floors, against the outer wall of a house, may be supplied and emptied from an outside shaft and vault. The earth is hoisted and discharged into the reservoirs from without. The boxes which receive the deposits and their covering of earth are tilted outward into the shafts.

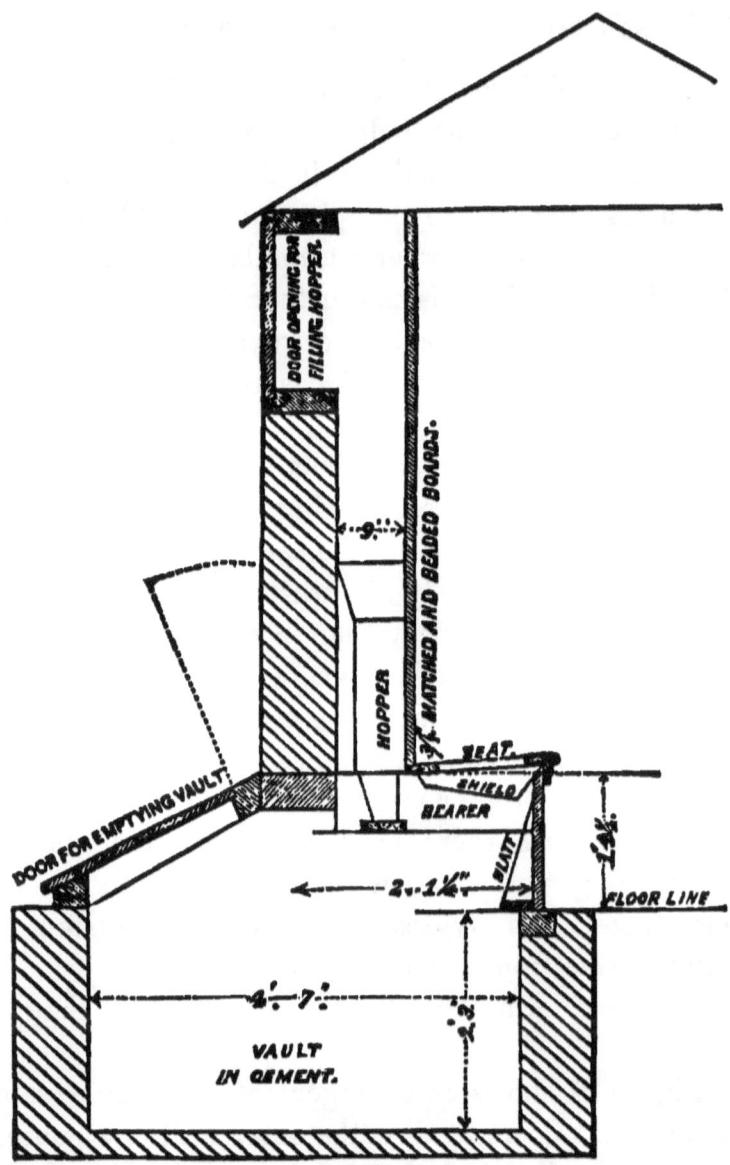

Figure 20. — Section of vaulted privy. To be supplied and emptied from the rear.

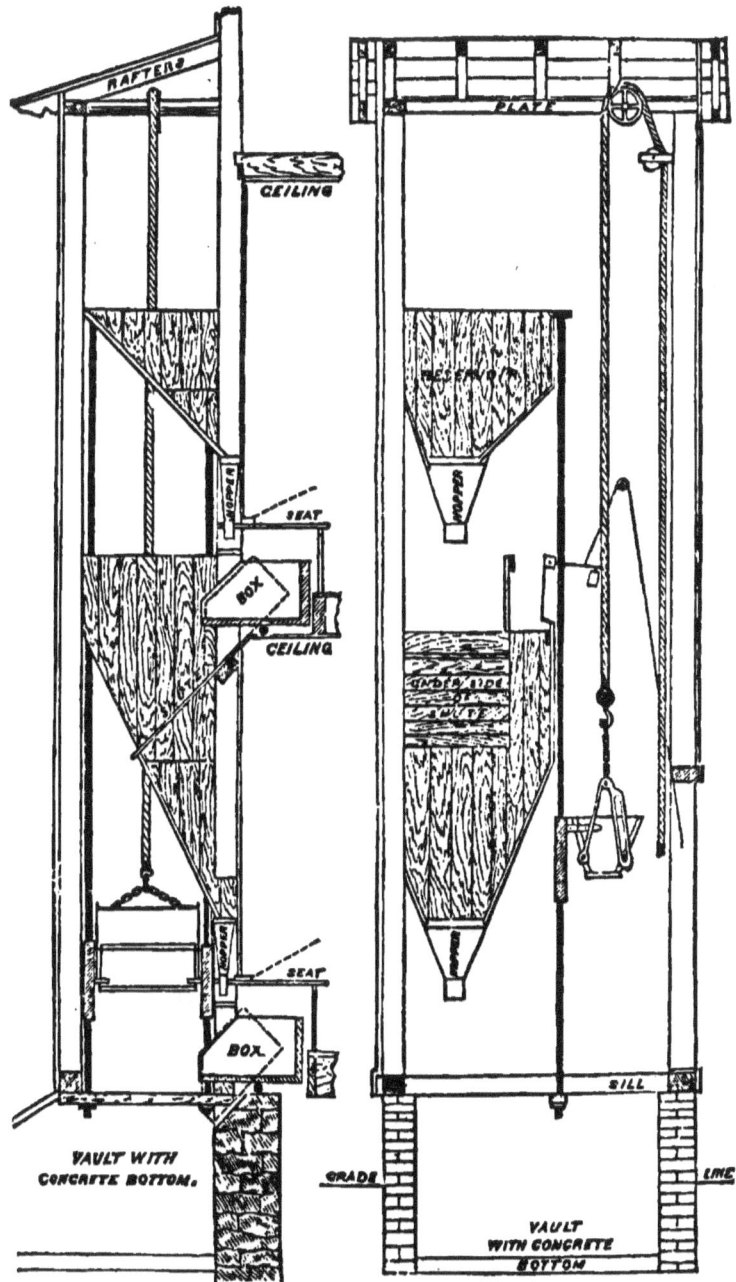

Figure 21.—Plan of closets on two floors, with hoist and dump.

The earth-closets at Fort Adams (Newport), erected for the use of the garrison, are constructed as shown in Figure 22. The deposits are received on the concreted floor of the casemate in which the closets are placed. The same plan — with divisions between the seats — is well adapted for the use of schools, asylums, etc.

Mr. Netten Radcliffe, in 1869, made a careful investigation of the earth-closet system in practical use in various parts of England, in connection with Dr. Buchanan. Their comprehensive and satisfactory report was published in connection with the Twelfth Report of Mr. John Simon, medical officer of the Privy Council.

In 1874, Mr. Netten Radcliffe alone made an examination of the privy system generally, and again inquired into the condition of the earth-system, — communicating the result of his investigations in connection with Mr. Simon's Report of that year. After this wide observation, he says, without qualification: "As a means of abating excrement nuisances the dry-earth system is of the utmost value." And again : " Of the value of dry earth as a means of abating excrement-nuisance, no question, I presume, now exists; and its application in detail to this purpose has been facilitated to the utmost by the ingenious mechanical arrangements devised and patented by Mr. Moule and Mr. Girdlestone (the engineer of Moule's Earth-Closet Company). These arrangements, which provide for proper charges of dry earth being thrown upon the deposited excre-

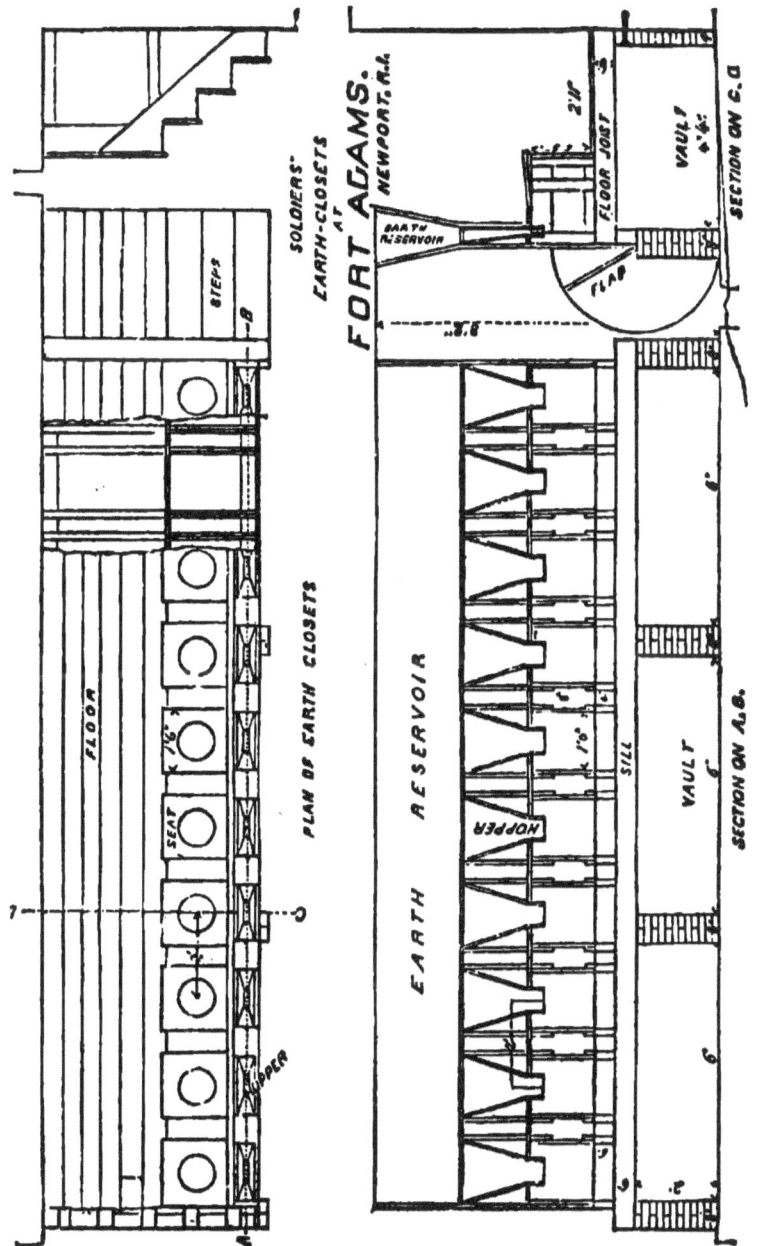

Figure 22.

ment, admit of ready adoption of the system in houses, schools, and other institutions.

"Since Dr. Buchanan's inquiry, the system in its integrity has been adopted in many mansions and on numerous estates, as well as in not a few public and private institutions. The wider experience of its use under these circumstances does not differ in result from that which has already been recorded by Dr. Buchanan, and it would serve no useful purpose to enter into a detailed examination here of the different instances which came under observation during this inquiry. So far as my observation went, wherever the system had been diligently applied and carried out, and due supervision over its working had been maintained, there its success in the abatement of nuisance from and the disposal of excrement had been assured. Where the system had been adopted without due regard to the amount and kind of labor at disposal and the amount of supervision which would be secured, there it had failed, under like circumstances. . . . .

"At Sinningróve, a village on the sea-coast, at the foot of the Cleveland Hills, and adjoining the Lofthouse iron-ore mines, I saw sixty-six earth-closets in operation. The mines are the property of Messrs. Pease, of Darlington, and the earth-closets had been introduced at the suggestion of Mr. France the manager. The closets, of which the mechanism had been constructed by Moule's Earth-Closet Company, were in excellent order, and the earth supplied to them, a clayey soil obtained from the foundation of buildings, had been carefully dried in a proper kiln erected for the purpose. A few of the closets had lever-seats, in others the earth was cast from the hopper upon the excrement by a handle acting upon a simply arranged "chucker." A man was detailed to prepare the earth and keep the hoppers supplied; and an arrangement had been made with neighboring farmers to remove the contents of the closets once in every three weeks. No difficulty had been experienced in making this arrangement; indeed, the farmers, it was told me, very gladly undertook the task for the value of the manure, and further, they had agreed to supply earth for the use of the closets when that obtained from new buildings failed. An inspection of the closets showed that the users had not habituated themselves to putting the mechanism in action after use, and that in consequence in some, although the hoppers were full of earth, the excrement was uncovered. This, however, appeared to have arisen rather from an oversight in the management than from any indisposition on the part of the cottagers to use the closets properly. It had been too readily assumed that the population for whom the closets were designed would take to their use without some instructional supervision. The advantages of the closet, as compared with the old-fashioned midden closet were, however, so obvious, even in the state that I saw them, that Mr. Frarce was about to introduce two hundred in a new mining village then in process of being built on the hills above Sinningrove."

He also visited some large collieries which had been supplied with apparatus by the Moule company. These are supplied with surface earth which is dried only under a shed, where it is occasionally turned over. These closets were found to be in a satisfactory condition.

"The several cottagers to whom I spoke on the subject were, with one exception, enthusiastic in their preference for the earth-closet as compared with the old midden closet, and more than one spoke of its greater decency, and of the influence of this upon the habits of growing children."

At Hereford, as in other cases investigated, it seemed sufficient to gather surface soil in dry weather, store it in suitable places, and use it without further artificial drying.

Mr. Radcliffe concludes:—

"It must not be too hastily assumed that the very fact of no local authority having adopted, of its own motion, the dry-earth system during the several years it has been before the public, is decisive against its adaptability to public requirements as to excrement disposal. The truth is, that only now does such a local sanitary organization exist as would admit of its application in those villages and towns where presumably the system is best fitted for operation. Before the Public Health Act, 1872, the sanitary organization of rural districts and of many small towns was too incomplete to give any reasonable hope of the efficient working of a system, whether the dry earth or any other, which required careful and systematic supervision and management. Since the passing of that act, an organization, fitted to these ends, has either been established, or is in progress of establishment, in every part of the kingdom. It is, perhaps, even more necessary now than when Dr. Buchanan reported, that sanitary authorities, in examining the sanitary requirements of their districts, should have under their consideration the dry-earth system, among other systems of dealing with excrement nuisances.

" I have already mentioned the great value assigned to the earth-closet manure by certain gentlemen, who are well acquainted with its practical use. This opinion, held also when Dr. Buchanan made his inquiry, has undergone no change, but has been confirmed by the five years' additional experience since that inquiry took place. On the other hand, Drs. Gilbert and Voelcker, studying the question chemically, have shown that the earth-closet manure after it has been charged twice, or even thrice, with excrement, is no richer than good garden mold.

" Mr. Walters, as I have stated, gets £6 a ton for the manure retailed in small quantities, and I may add, that he believes this sum fairly represents the value of the material. Dr. Voelcker estimates the value of the compost after it has been charged five times with excrement, at 7s. 6d. per ton. I cannot pretend to reconcile the differences; I merely state the facts. But it may be observed that the chemical estimate of the value of earth-closet manure, does not disprove the sanitary value of the dry-earth system, but, so far as it may be the true index of value, only tends to show that its economical adaptation must be limited to cottages and small towns, where the cost of providing and drying the earth, and distribution of the manure, will be of the smallest.

"On this question the Committee [1] on the Treatment and Utilization of Sewage, appointed by the British Association for the Advancement of Science, has said, as to houses and villages (again looking at the value as a matter to be estimated by chemical analysis), that the dry-earth system "might be even economical where the earth for preparation and absorption, and the land for utilization, are in close proximity."

"Without desiring to underrate the commercial aspects of the question, it appears to me, that it is the economical aspect in the sense of obtaining an unquestioned good at the least cost, which has place here. If the value of a method of excrement disposal is to be estimated by its profitableness as a pecuniary investment, rather than by its hygienic success, all measures at present in use in this country would have to be condemned. From the former standpoint, the best, perhaps, that can yet be said of the completest of these is, that it is the least costly. From the sanitary standpoint it is unfortunate, although quite explicable, that the pro-

[1] Report of 1872, p. 188. Drs. Gilbert and Voelcker were both members of this committee.

moters of the dry-earth system should have rested its advantages so largely upon its presumed results for agricultural purposes. Their experience must, however, be taken as showing that there are certain conditions of use of earth-closet manure which justify their encomiums of it as a manure; and there is no sufficient reason to believe that a multiplication of like experience would lead to different results. But adopting the chemical estimate of the value of earth-closet manure, it still leaves the question in this not particularly unfavorable position, namely, that the dry-earth system is perhaps the only method of excrement-disposal at present practiced in this kingdom, which wholly, or almost wholly, would probably cover the cost of working, if it were judiciously put in operation within suitable districts."

. . . .

" I now find myself in a position to state, with some approach to accuracy, the way in which the earth system may be worked, as well as its approximate cost and produce. I need not here consider the case of public institutions, or very small villages, as the instances quoted sufficiently illustrate the operation of the system there. But for my present purpose I begin with the case of a village population of 1,000 persons already provided with the ordinary arrangement of outside privies and cess-pools. People making use of closets as receptacles for all stools and urine from every inhabitant, may be taken to use them on an average three times a day each, and to require for each use 1½ lbs. of dry earth. This gives 4,500 lbs., or two tons, as the daily quantity of earth required for the population. The amount that would accumulate in the closet pits, and which would need to be removed about four times a year, would be larger than this, by the bulk of the stools, and of such portion of urine as did not evaporate, but without reckoning increase on this score, the quantity of manure produced may be reckoned at the same quantity of two tons a day.

"I assume that, after owners of property have paid the original cost of providing earth-closets according to the scheme of the local authority, all supply and maintenance of them should be the function of that authority. The cost to owners would vary (1) according to the adaptability of the existing arrangements, and (2) according to the character of the earth arrangements to be required. The latter may either consist, as at Lancaster, in a single daily application of earth to the closets, or much preferably, as at Halton, in an arrangement for the mechanical delivery of earth after each use of the closet. In this latter case, an average outlay for structural alterations and machinery of some £3 or £4 might be required in respect of each closet."

He estimates the capital needed for the original plant, at £250; and the weekly outlay for earth and labor, at £4 15s. The annual cost, including interest on plant, will be £260. The product will be 730 tons of manure, costing seven shillings per ton.

"The extension of this scheme, beyond the village

of 1,000 people, to larger towns, appears to be essentially a question of multiplication, with these differences: on the one hand, an organization on a large scale can commonly be had more cheaply than one on a small scale, and in this way and by its compactness the town has the advantage over the village; on the other hand, labor is dearer in towns, and towns often have their closets so arranged that it is difficult, without much cost, to adapt them to the earth system, and thus the village has advantage over the town. Further, in towns which must necessarily be supplied with sewers for the purpose of drying the soil, and for removing rain-fall and house slops, the question arises whether it may not be more advantageous to throw all foul matters together into these sewers. I do not propose to discuss the relative merits of a water-closet system and of an earth-closet system; this must depend upon a variety of considerations proper to each particular place. In a locality where sewage can be cheaply delivered upon suitably situated land, where the amount of sewage dilution is such as fits it for the particular crops that are marketable, where the irrigable land is of such extent and quality as effectually to remove the manurial constituents of sewage, and to allow of the effluent water passing off in sufficient purity, in short where sewage irrigation can be effected with profit to the people and safety to the health of themselves and their neighbors, I should anticipate a preference for a system of water carriage for the excrement of the place. But for populations where these conditions may not be attainable, or where experience may show greater profit realizable from solid manure, I should suppose that the earth system would find advocates in preference to the water system; and it is impossible to ignore the fact that many large English towns do not regard the water-closet system as suited to all their particular wants, nor irrigation as being a remedy certainly suitable to their particular sewerage difficulties. I refer, of course, to towns which, although possessed of a system of sewers, nevertheless retain their excrement in middens or cess-pools, deliberately avoiding water-closets as not affording them the certainty of advantage which they need to have before they enter upon expensive new constructions. By the authorities of such towns the earth-system will especially deserve consideration as promising them the means of making harmless their retained excrement by a system readily, perhaps, adaptable to their present privy construction, and not involving in its introduction a new kind of difficulty.

" The present inquiry has led me to conclusions as to the hygienic advantages of the dry-earth system similar to those arrived at by Dr. Buchanan in 1869, and I adopt mainly his words in stating them.

" 1. The earth-closet intelligently managed, furnishes a means of disposing of excrement without nuisance, and apparently without detriment to health.

" 2. In communities, the earth-closet system requires to be managed by the authority of the place, and in limited communities it will probably pay at least the expenses of its management.

"3. In the poorer class of houses, where supervision of any closet arrangements is indispensable, the adoption of the earth system offers especial advantages.

"4. The earth system of excrement removal does not supersede the necessity for an independent means of removing slops, rain water, and soil water.

"6. As compared with the water-closet, the earth-closet has these advantages: 'It is cheaper in original cost; it requires less repair; it is not injured by frost; it is not damaged by improper substances being thrown down it; and it very greatly reduces the quantity of water required by each household.'"

A large manufacturer in Lancashire thus states his experiences: —

"Having for some years previously felt much interest in sanitary affairs, I decided, in 1870 (in no small degree, by reason of the very favorable notice of it by Dr. Buchanan, in the twelfth report of the medical officer of the Privy Council), to try Moule's earth-closet. I pulled out a good water-closet and substituted an earth-closet. A very short experience caused me to do away with every other form of closet I had, and adopt Moule's throughout. This was all completed some time ago, and there are now twenty-six at work; nine at a cotton factory, three at my own house, and the rest connected with cottages. It has been objected by some to Moule's closet that mechanical contrivances are apt to get out of order. That is true, but from the first up to this time, I have had nothing whatever that has gone wrong. The closets at the factory and one closet at my house (inside and up-stairs), are fed with clay, the others with sifted coal ashes, so that ten out of the twenty-six are fed with clay, and the produce of these we sell; the ash manure I use myself."

THE GOUX EARTH-TUB SYSTEM.

I have had no personal knowledge of this system, and take the following from English sources.

The largest trial that has been made with the Goux system is in the town of Halifax, in England, where in May, 1874, Netten Radcliffe found 2,573 closets so arranged. In his report he describes the system as follows:—

"A detailed examination of the working of the system in Halifax showed, as a rule, a less degree of offensiveness to the eye than is commonly observed in the simple pail system.

"The Goux system is a pail system of which the peculiarity consists in a certain preparation of the pails, and in a particular mode of manufacture of the excrement into manure, and utilization of the dry house refuse generally. I am here concerned with so much of the system only as relates to the abatement of excrement nuisance in the vicinity of dwellings. The pail used in the Goux system is preferably of wood, of oval form, and measures 24 by 19 inches, and 16 inches deep. It is prepared for use by being lined at the sides and bottom three or four inches thick, with various refuse matters, used as absorbents. These matters may be (to quote from a trade circular), 'chaff, chopped or broken straw, damaged or refuse hay, coarse grasses, moor grass, dry street sweepings, dry horse-dung, and litter, sweepings of markets, hay and straw lofts, refuse wool and hair, wool, shoddy, vraic or seaweed, charcoal dust, dry peat, dry ferns, spent dye woods, coal ashes, etc., any, or all of these, or their equivalents, to be mixed in such proportions as may be most convenient, together with a small percentage of sulphate of iron, or sulphate of lime. At Halifax, the materials used for lining the tubs are the waste arising from the manufacture of worsted, cotton and flax, and old manure which has become dry and fallen to powder. To these materials a little sulphate of lime is added. The pails are lined with the assistance of a mold.

" The lining of the pail is designed to absorb the urine and other liquid which may pass into the pail, and so tend to keep the excrement drier and delay its decomposition; but the absorption appeared to me to be trivial in pails used by women and children. Widely different degrees of sloppiness existed, obviously dependent upon differences in the families using the pails; but the extent of sloppiness noticed in Salford, in 1869, was rarely observed in Halifax, greater care being apparently taken in the latter town to avoid the emptying of chamber utensils into the pails. Probably the more regular locking of the doors of the closets, which is practiced in Halifax, contributes not a little to the exclusion of the contents of chamber utensils from the pails, less trouble being experienced in casting them into the yard drain. At any rate the aspect of the lined pails in use in Halifax generally, was less offensive to the eye than that of the simple pail, and the casting down of a portion of the lining, as I noticed in several instances, sufficed effectually to hide the offense and diminish the odor from the pail.

"The sanitary advantages gained from the introduction of a pail system, such as Goux's, as compared with the midden system, in Halifax, cannot well be overrated. The specification for the reconstruction of privies on the Goux system, necessarily provides for the filling up of middensteads; and the suppression of those receptacles is an initial requisite of sound privy administration. . . . .

"Some needless carelessness occurs in lining the pails, in their removal and cleansing, as well as in the cleansing of the night-soil vans, from want of special supervision of the working of the system by officers of the corporation. I noticed in the course of my inspection, pails imperfectly lined, and some not lined at all; pails which had overflowed from neglect to remove them at the proper time; littering of ash-place in the removal of the ashes; some splashing in van and leakage from van into street; and unsatisfactory arrangements for cleansing the vans at the wharf whence the pails are dispatched to the manure works. These defects, insignificant as compared with the advantages which those parts of the town derive from the system where it has been introduced, but exciting prejudice against it, entirely arise from the want of such special supervision as the corporation should exercise over it."

The Rochdale Corporation made an experiment of several months with the Goux system, and set it aside as being less simple than the small pail system, which they found satisfactory.

Dr. Alfred Haviland, Medical Officer of Health to the Rural Sanitary Authorities in the counties of Northampton, Leicester, and Bucks, publishes in the "Sanitary Record" of September 25, 1875, a very strong indorsement of the Goux system, from which the following is quoted : —

" Having been favorably impressed with the principle of the Goux system, which I saw in operation at Aldershot in October last, I was anxious to investigate it further, and for this purpose visited Halifax in March last, in company with Dr. Goldie, the medical officer of health for Leeds. Halifax, at the census 1871, contained 13,970 houses and 65,510 persons. It is situated between high hills, and its site is so irregular, that many of its main streets and roads are particularly ill-adapted for the draught of heavily laden vehicles, so that scavenging on a large scale is performed under greater difficulties than perhaps in any other town in England.

THE DRY CONSERVANCY SYSTEM. 269

"We were first taken to the manure shed, where the sewage is taken direct from the houses: a building 145 feet long and 74 feet wide, roofed with wood and felt, with open rafters, it contained 1,500 cart loads of manure, from the fresh contents of the tubs to the ripe manure fit for the field. There was no real sewage smell, although ammonia was perceptible, not upleasantly, however ; the odor of the atmosphere resembled that of a well-kept cow-shed.

"In Halifax, the lining of the tubs is shoddy, at Aldershot, stable litter. A lad lined six tubs and prepared them for use in two minutes.

"The company supply in the borough 3,020 closets with tubs — a fifth more is kept in reserve. About 550 tubs are emptied daily into the shed, twenty-one being a load. Scavenging begins at 7 A. M. and ends at 5.30 P. M. Occasional complaints have been made of the smell of the tubs, but these have only occurred when the emptying has been neglected, and the tubs allowed to get too full, as was the case during the severe parts of this year, when the horses could hardly stand on the slippery hills of Halifax, and the scavenging could not proceed at the usual rate. The tubs are emptied once a week.

"The sewage is emptied directly into the shed ; it heats like stable dung. At the depot the manure sells at eleven shillings per ton, and two shillings extra for carting to the station, which is about a mile off. It is then conveyed in open trucks by railway to its destination.

"Only one complaint has been made as to the smell arising from these trucks, which are loaded at the ordinary goods station.

"In the work twenty-one men are employed and eighteen horses, ten belonging to the company and eight hired. As to the health of the men employed, not one had lost a day's work from sickness for the last two years. The Goux system was established in Halifax in January, 1870."

Netten Radcliffe says in his report of inspection of the system at Salford, —

"We inspected many pail-closets used by single families, and others used by several families, or by the inmates of a common lodging-house. In every instance where a pail had been in use over two or three days, the capacity of absorption of the liquid dejections, claimed by the patentee for the absorbent material, had been exceeded; and whenever a pail had been four or five days or a week in use, it was filled to the extent of two thirds or more of its cavity, with liquid dejections, in which the solid excrement was floating. The contents, in fact, differed nowise in aspect, except in the cases where a portion of the dyewood lining had broken down and fallen into the liquid, from what we should have expected if a simple unprepared pail had been used. It was suggested that a part of the sloppiness of the pails probably depended upon the fact of chamber pots having been emptied into them; but although the regulations for their use permits this to be done, we did not always find on inquiry that even this source of wetness had been in operation."

THE DRY ASH-CLOSET.

Although this system is properly a branch of Dry Conservancy, it may (as being usually applied to some form of vault closet) be more appropriately described in the next chapter.

NOTE, 2d edition.—Experience and careful analysis have recently shown that in properly prepared earth the entire organic matter of both urine and fæces is completely destroyed by the oxidation which is always active in aerated porous material. The same earth may be used over and over again, for an indefinite time, if it is allowed to lie under cover for a month or six weeks after each use. This vastly simplifies the question of earth-supply.

This matter is fully discussed in my *Sanitary Condition of City and Country Dwelling-Houses.* New York: D. Van Nostrand.

CHAPTER VIII.

VAULTS AND PRIVIES.

ANY perfect sanitary system would probably require the entire abolition of all cess-pools and vaults deep in the ground, and of all receptacles of every sort where the matters to be treated are allowed to accumulate in considerable quantities. There is no doubt that could we have our wish, it would be best in every case to make an advantageous application of the water, the pneumatic, or the dry conservancy system.

But the cases are very numerous where public opinion is far from being sufficiently educated in sanitary matters, where the powers of the sanitary authorities of the town, or village, are far too limited and where too much general indifference exists, for anything like a radical reform to be undertaken with the hope of success. In such cases all that can be attempted is to reach such a modification of the methods now in use as shall render them at least much less offensive and dangerous than they now are.

This subject has had great attention from local sanitary authorities in many towns and districts of

VAULTS AND PRIVIES. 273

England, and the investigations concerning it have been most painstaking and valuable.

It seems evident that there are but three roads of escape from the annoyances now existing, short of a more thorough system, whose adoption is, in so many cases impracticable. These are: The pail system, with frequent removal; the tight vault, to be emptied as occasion requires, with movable pneumatic apparatus; and the ash-pit system, with frequent removal.

If the Goux system is to be considered worthy of public confidence and general adoption in any town (which seems doubtful), its management should be subject to the directions and restrictions given below with reference to the pail system.

THE PAIL SYSTEM.

This, which is also called the Rochdale system, consists in the use, beneath the seat, of a tub (usually the half of a petroleum barrel) to receive dejecta unmixed with any absorbent, — the tub or pail to be removed at frequent intervals. During the removal a tight cover is used, and the pails are carried in a covered wagon to a depot where the excrement is mixed with ashes and sold as manure. In all systems where a removable receptacle is used in town closets the round form is better than the square, as being more easily kept clean, and the size should be not only such as to be easily handled by one man, but such as not to admit of remaining too long without cleansing. These receptacles should not simply be emptied into a scavenger's cart, but should be taken to the depot to be cleansed and well aired, their places in the closet being supplied with fresh vessels.

The Sanitary Committee of the Rochdale Corporation say, with reference to the small pail system, that "the essential condition of the trial, *frequency of removal*, had been secured; that the system of removal had been thoroughly approved by all who had had experience of it; and that it had not failed under most varied conditions, having proved equally efficacious in the highly-rented house with its own closet, in the lodging-house where great numbers were accommodated, and in the factory and workshop. In the subsequent manufacture of the dejections and ashes into a salable manure, the committee concluded that the Goux system was less advantageous than the use of tubs without absorbent linings."

The pail system has been less successful in Leeds than in Rochdale, but evidently only for want of proper attention and sufficiently frequent removal, which indicates the leading objection to any such system as this under any but the most careful management. Well managed, any of the removal systems will be satisfactory, while none will bear neglect among the poor class of a population much better than will the ordinary water-closet system. The pail closet is gaining favor so rapidly in Biringham, in the estimation of property holders, that the means thus far provided for its extension are taxed to the utmost.

PNEUMATIC EMPTYING.

There is much to be said in favor of tightly cemented vaults to be emptied by portable pneumatic apparatus; and this emptying process, as applied to common vaults, is achieving a success which — when we consider the still-prevailing horrors of hand-emptying — is well deserved.

But this system is open to the grave objections, that in practice the vault would often be anything but tight, and would in such cases have all the defects of the common privy; and that even when tight, its purpose would be to retain in the vicinity of the dwellings, and in a state of putrefaction which must always endanger health, matters which it should be our greatest aim to remove at once or to retain (as with the earth system) in an innoxious condition. These objections are so grave that they should suffice to condemn the whole process, save as a make-shift for use so long as common privy vaults are tolerated at all.

THE ASH-CLOSET.

The ash-closet which has come into use (and into great favor) in several large towns in England, is usually intended to be emptied at frequent intervals. Its best development seems to be in the town of Hull.

The daily removal recommended by Doctors Radcliffe and Buchanan in 1869 has, however, not been carried out, weekly removal being thought to be all that was practicable. These gentlemen say: —

"In the present imperfect state of our knowledge of the conditions under which fæcal diseases spread, we do not feel ourselves entitled to say at what time, after being passed, dejections are or may (under various external circumstances) become dangerous to health. We cannot say this either in regard of healthy excrement, or of that passed from persons affected with disease, specific or other, but we think it may probably be taken as sufficiently true for practical purposes that there is little chance of mischief from the storage of excrement *for a day*, even though along with healthy excrement that of persons affected, for example, by enteric fever should, without proper disinfection, chance occasionally to be included. We propose, then, to regard *complete removal of all excrement within a day* as practically constituting safety in the case where excrement is unmixed, or is only mixed with ashes."[1]

Dr. Radcliffe advises in his later report: "In all forms of fixed closets the foremost condition, — the one to which all other considerations should yield, is the *frequency of removal of deposited excrements.*"

For the information of town authorities who may contemplate adopting a system of frequent removal, it may be interesting to repeat the following from Mr. Radcliffe's report of the regulations in force in the town of Hull, England.

[1] Meaning, doubtless, ash-bin refuse.

" For the purposes of 'night soil collection' the borough is divided into forty-eight districts, each containing from three hundred to seven hundred houses, the total number of houses enumerated for the purposes of this collection in 1873 being 30,977. The collection is carried out by the sanitary authority through the agency of contractors. As a rule, each district is let out to a separate contractor, and no contractor is allowed to undertake more than two districts. The smaller districts are so arranged that the collection may be carried out by any one who has the command of a horse and cart, and who can have the assistance of a boy or two. With a view of obviating undue combination among the contractors, and diminishing the evil effect of strikes among the men, the contracts are so timed in the letting that only eight or nine can fall vacant together. The contractor, in addition to receiving the material he collects, and which he sells for such profit as he can obtain, is paid by the sanitary authority from 2s. to 3s. yearly for each house in his district. The sanitary authority provides places of deposit (four in number) where the contractor can store the collected material until disposed of; and he is subject to penalty if he should deposit such material elsewhere without permission of the inspector of nuisances, in writing. The contractor is required by the terms of his contract to collect and remove *at least once a week*, all night soil, offal, dry and liquid filth, dust, paper, and other refuse of every description, from all premises, middensteads, ash-pits, dust boxes, cellars, or other places used for such refuse, attached to all houses, shops, warehouses, yards, and other premises within his district, with the exception of trade refuse exceeding in quantity three cubic feet in any one week, and all contents of cess-pools, blood, manure, and filth from slaughter houses, ashes from furnaces, and refuse from manufacturing processes. The work of collection and removal is to be executed on week days, from the beginning of March to the end of October, between the hours of 5 and 8.30 A. M. and from the beginning of November to the end of February, between the hours of 6 and 9.30 A. M., and all carts employed in the work are to be clear of the streets and public thoroughfares, on their way to the depots, before 9 A. M., within the former period, and before 10 A. M., within the latter period. Further, the contractor is required to use water-tight and properly covered carts."

The arrangement of the Hull privy is extremely simple as indicated in the accompanying illustration.

" The space under the seat forms the entire receptacle for all the ashes, refuse, and excrement of the house, and is built of bricks in cement, with a bottom of brick or flag, sloping from the level of the paved floor in front to a little below the ground level at the back, and forming only a very shallow pit. Into this space, through the hole in the privy seat, all dry refuse is thrown. The front of the midden space is formed by the front board of the closet, which is made moveable, to give the scavenger access to the pit. There is no drain to it, as rain is excluded and slops are in practice thrown down the drains. The ashes are usually sufficient in quantity to soak up all moisture passing into the pit, and the contents are almost invariably dry, and are removed by a spade without difficulty."

Simple as this construction is, its adoption in our northern towns would require some provision should be made which should prevent its receptacle being cracked by heaving from frost, but if properly constructed, and if frequently cleansed under efficient supervision, it would certainly be a very great improvement on the system at present in general use.

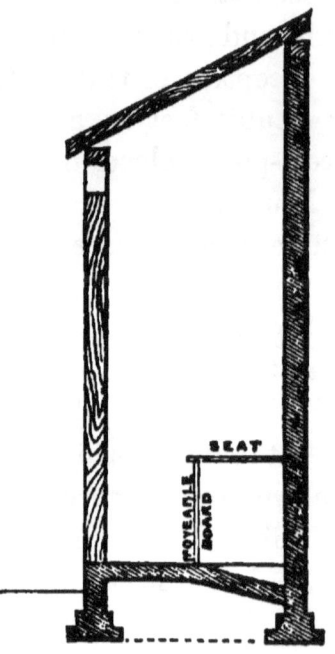

Figure 23. — Section of privy.

The instances in Hull, in which the arrangement was found to be unsatisfactory, seem to have been due to one or other of the following conditions: —

"1. Deteriorated, or original imperfect construction of the walls of the privy pit, leading to retention of portions, and perhaps to some soakage of decomposing filth.

"2. Careless casting of slops into the privy pit,

facilitating decomposition of the contents, and soakage of the wood-work.

"3. Want of adaptation of the scavenging to the needs of particular localities and their inhabitants. In the localities occupied by the most impoverished and degraded of the population the privies were overflowing with filth, and most offensive. This arose mainly from the insufficiency of the scavenging. Designed to meet the requirements of a single family only, the Hull privy cannot be used by several families without being productive of nuisance, except on condition of a more frequent removal of its contents than once a week. A daily removal is necessary under these circumstances; and as to orderliness of the privy, in those cases where a single family cannot be made responsible for it, this will not as a rule be secured unless the sanitary authority itself undertakes the duty of maintaining it."

Netten Radcliffe made a careful examination of the Dry-Ash system in Manchester, where 6,000 such privies were already in use, and thus reports : —

" In the series of inspections I made with reference to the working of this new system, I had occasion first to observe the contrast as to nuisance between the dry-ash closet and the old midden closet. In several streets where the process of reconstruction had been only partially completed, it was possible to compare the old and new privy arrangements in contiguous premises. It was the contrast between open, big, uncleanable cavities, containing a greater or less amount of decomposing fæcal matter, and emitting a horrible, penetrating odor, and small receptacles emitting hardly any appreciable smell, even with the nose above the *privy seat, and admitting of thorough cleansing. Most significant testimony was given to the benefit of the change by some householders. Many houses in Manchester are built in parallel rows, a back passage running between the rows, and each house having a small yard in the rear in which the privy is placed. Since the reconstruction of the privies, '*it has been possible to open the back windows of the houses.*' The change, moreover, has affected beneficially the value of cottage property, and tenants are quite willing to give 3*d.* more rent weekly since the reconstruction of the privies, for the gain in decency and comfort. Soakage of excremental matter into the soil, and its passage into and accumulation in drains is, of course, obviated by the reconstruction, and the smaller space occupied by the new closet is not an unimportant matter. The removal of the excrement pail is, with the most ordinary care, free from offensiveness, and if commonly conducted as I saw the operation, it may well be executed during the day-time, and the abomination of night scavenging done away with. The use of the cinder sifters has been adopted by householders with a readiness which proves how accurate the corporation was in depending upon their coöperation in the working of the scheme. The high price of coal during the last two years has contributed to this good result, from the value of the cinders in economizing its use. It is found, also, that a class of the population, commonly believed to be unmanageable in regard to any niceties of arrangement for excrement disposal, have rapidly appreciated the advantages of the new closet, and taken to the use of the cinder-sifter. In other words, it has been found that habits of decency and order in the particular matters referred to have been largely developed with the opportunities for such decency and order. Among the lowest class, occupying sub-let houses, and having privies used by families in common, it will, however, probably be found necessary to adopt some special supervision, and to remove the excrement and dry house refuse daily."

Where these closets are in use the instructions of the sanitary authorities require the inspector of nuisances to report as nuisances all closets in which the due covering with ashes or earth is neglected.

TUMBLER AND TROUGH CLOSETS.

These are closets for the use of large numbers of persons (as in factories), where there is no other objection to the water-system than the liability of the usual apparatus to get out of repair.

They each have a trough under the seat, through which water is either kept running, or in which it stands to a certain depth, to be let off from time to time. The "tumbler" is similar to that described (page 167, Figure 7) under the head of "Flushing Sewers."

The measurements of the tumbler for a closet may be: —

Length of tumbler, at top	3 feet.
Length of tumbler at bottom	1 foot 10 inches.
Width of tumbler at bottom	1 foot 7¾ inches.
Depth of tumbler at back	1 foot ¾ inch.

Trunnions are 6¼ inches from top, and 1 foot 1¾ inches from back.

In Leeds, the use of the tumbler-closet has not been extended during the past three years, it being considered wasteful of water, and difficult to keep in order.

TROUGH CLOSET.

In Liverpool, where the trough-closet, flushed with water, is in quite extended use, Dr. Trench stated that, in 1868, when an epidemic of enteric fever was prevailing, "The only localities which seemed exempt from it were the places occupied by the poor, in which we had removed all the privies and made trough closets."

CHAPTER IX.

LIERNUR'S PNEUMATIC SYSTEM OF SEWERAGE.

THE important problem of town sewerage seems to be seeking its solution by the aid of all the natural elements. Water and earth have had their trials and have been more or less successful, and now an ingenious Dutch engineer has called air into requisition, and promises to solve all the difficulties which have been but partially overcome by previous systems.

Captain Charles T. Liernur, of Holland, a military and civil engineer of much experience (long a railroad engineer in America), has devised a system for which he claims great results, and which, theoretically at least, seems to possess advantages far beyond those of any other that has been applied to densely populated town areas. This system has, as yet, been too incompletely tested, and some of its important supplementary details have been too little experimented with, for one to say definitely that it is an assured success which is entirely to drive from the field the water sewerage now in such general use ; but its claims are set forth with such positive assurances of merit, and its various parts seem to have been so well considered, that it is worthy of more than passing notice as merely a curious mechanical contrivance.

As every important invention in connection with the removal of the fæcal matter of towns should be approached in a hopeful spirit, and encouraged by the fullest opportunity for its development, it will be best first to state what are, and what are to be, the mechanical details of Liernur's process, and what its adherents believe that it will accomplish.

The initial principle of the system lies in the suction to a central public reservoir of the accumulation of fæcal material deposited in receptacles at separate houses, these being connected with this reservoir by air-tight pipes. The reservoir being exhausted of its air, the accumulations are drawn toward it by pneumatic pressure. No matter how large may be the area occupied by the sewered houses, each district has its central reservoir, and these reservoirs are in turn and in like manner themselves discharged into a main vacuum chamber at convenient point, being connected with this by a similar system of pneumatic pipes. The deposits at each house are first removed to central receptacles in their districts, and the whole mass is by a second or even by a third operation drawn to the main depot, where it is to be disposed of according to the requirements of the conditions of health, and most conveniently for agricultural use.

The invention has grown gradually from small beginnings, and it has been in one or two instances applied over large areas with very satisfactory results. As the system in a town of even the largest size is merely an aggregation of smaller systems, to describe one of these latter will suffice for an understanding of its principles.

We will assume, then, a level town area of from one hundred to one hundred and fifty houses of medium size. In the centre of this area, in the middle of a street, and far enough below the surface to be secured against frost, there is sunk an air-tight iron reservoir having two openings at its surface, to either of which an air-pump connection, or the connecting pipe of an exhausted receiver may be attached. The air-pump attachment, used to create a vacuum, opens into the top of the reservoir, while the attachment of the exhausted receiver, being intended to suck out the liquid contents, is connected with a pipe reaching nearly to the bottom.

When the air-pump is applied for the exhaustion of the air of the reservoir, it creates a partial vacuum, which extends through the whole series of pneumatic pipes connected with it, and the pressure of the air entering at the remote open ends of the pipes drives forward toward the vacuum-centre all of their liquid accumulations.

After the reservoir has become filled, the pipe reaching to its bottom is attached to the previously exhausted receiver, into which the liquid is drawn. Main pipes, under ground, running through the streets, or through the spaces between the backs of houses, and with branches to or under the houses themselves, allow the accumulations of the house closets to flow to the reservoir whenever a vacuum is established and is, by the opening of stop-cocks, brought to bear upon them. The closets of each house, which may be placed one over the other on the different stories, are connected with the branch pipe described, having a vertical or nearly vertical fall to the point of junction. When the cocks are opened, so that these branch pipes are brought into direct communication with the vacuum, every house pipe, being open at its upper end, becomes a source of pressure, and the air in seeking to fill the vacuum carries before it whatever matters may be accumulated within it. In the earliest introduction of the system, each house branch was supplied with a cock, so that after the reservoir had been exhausted of air, the opening of each of these, for a moment, caused the contents of its pipe to be thrown rapidly forward toward the street reservoir; but as there was no means of knowing the exact time needed for the emptying of the contents of each pipe, either there was necessarily incomplete work, or more air might be admitted than the work required. Later, there was substituted for these stop-cocks an arrangement of self-acting air-traps which entirely overcame the difficulty. These traps give equal barometric resistances, and by their aid the accumulations of each house, be they great or small, far or near, are discharged with absolute uniformity and regularity by the opening of a single cock in the main pipe with which the house branches are connected. These automatic traps, depending for their action on this equal barometric resistance, are not merely effective for the purpose for which they were intended: they are also interesting as a most ingenious and curious invention. Their action may be easily explained.

The accompanying diagram (Figure 24) shows two tumblers containing water. One is nearly filled and the other has but an inch of water at its bottom; the difference in height between the two levels of the water we will assume to be two inches. The barometric resistance (against suction) is greater, by the pressure due to a column of two inches of water, in the one than it is in the other. Into each of these two tumblers a glass tube is inserted, and the ends of both tubes are taken into the mouth at the same time. We will assume that the vertical height between the surface of the water in one of the tumblers and the mouth is four inches, and between the surface of the water in the other tumbler and the mouth is six inches; consequently in one case there is a col-

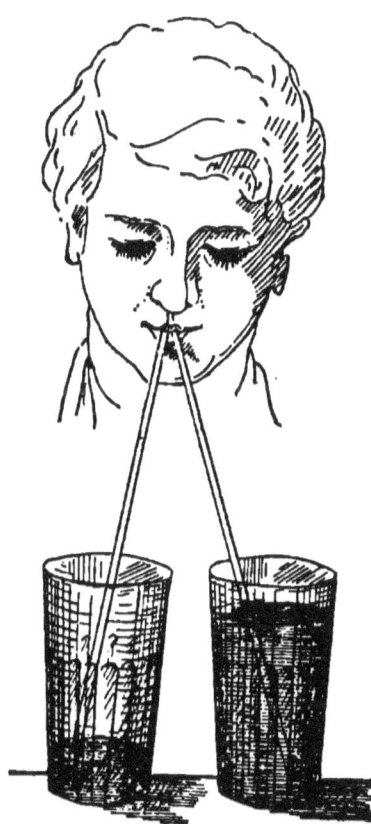
Figure 24.

umn of four inches of water to be lifted, and in the other a column of six inches. Now if one sucks very gently on both tubes, that is, if both are *slowly* exhausted by the same mouth, water will flow only from the tumbler which is the fuller, or from which the shorter column is to be lifted, until the level of its water is reduced to the level of that in the other tumbler; then, the height to be overcome being equal, there will be an equal flow from each tumbler until both are exhausted. No matter how much water there may be in one vessel nor how little in the other; if the same slow draft is made on both at the same time, the flow will always be entirely from the one standing at the higher level, and after the equilibrium is established there must be an absolute equality of level preserved until both are exhausted. The same effect will be observed if we experiment with a dozen tumblers, all having their contents at different elevations; that one in which the liquid stands at the highest level will be discharged first; when this reaches the level of the second, these two will be discharged together; when these descend to the level of the third, the three will deliver equally; and so on until the whole series, offering an equal resistance to an equal force, deliver their contents at the same rate.[1]

[1] The flow through the tubes must be so slow that the element of *friction* shall not interfere with its success. Practically, it is difficult to make the draught by the mouth sufficiently slow and steady for success with the small tubes required. With an air pump or compressed rubber ball it is easier to regulate the force, and the tubes may be larger.

Captain Liernur has applied this principle of barometric resistances to his pneumatic tubes by giving to each (for convenience, before it leaves the premises by which it is supplied) a break, or abrupt change in elevation, of say exactly one foot. It is necessary that there should be always a distinct fall, or inclination toward the direction of the flow of the pipe, so that its liquid contents may move forward without halting at any point to deposit silt, which might in time obstruct them. Practically, it is said to be best to give an inclination of one foot in a length of fifty feet. This for a minimum ; the maximum may be whatever circumstances require. In a level district all the pipes of the system may have this minimum inclination, but where the town is built on irregular surfaces one pipe may lie at this slight pitch, and the very next one may, without detriment, have an inclination of forty-five degrees or more. All tend toward the same central point, and may have more or less fall in that direction. But each pipe has its flow interrupted by the trap or vertical step referred to. Figure 25 shows two such pipes, leading from two different houses and delivering to the same street main: *a* is a pipe with a very steep inclination, and *b* is a pipe at the minimum

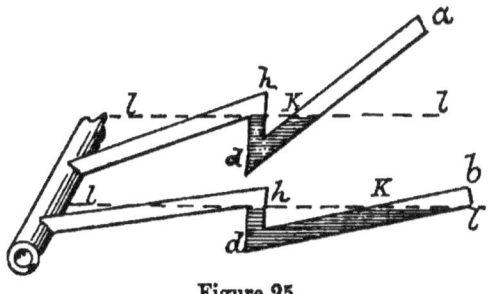

Figure 25.

inclination. The dotted lines *l l* show the height to which the liquid must rise in the pipes toward *a* and *b* before it can begin to flow over the high points *h*. If the production of either house is more than enough to fill the depression in the pipe below the dotted lines, any addition to the quantity will simply cause a discharge by gravitation over the angle *h*, and the liquid will flow on by its own force toward the reservoir. This flow will of course continue so long as there is an addition to the volume at the higher end, but the amount of liquid standing below the level of the dotted lines must always remain there until some artificial force is applied to move it. Now suppose the suction of a vacuum to be applied at the main pipe *c*. The pressure of the air is brought to bear on the surfaces of the liquid at the points *k*, forcing the whole mass forward over the high points *h*. The flow begins at the same instant in both pipes, but as there is a larger volume in the pipe having the more gradual (and longer) slope, and as the vertical descent of the two surfaces must be exactly the same, the amount flowing out of the pipe *b* will be greater than that flowing out of the pipe *a*, until *k* has descended to the lowest point *d*, when in both pipes there are equal columns to be overcome (from *h* to *d*), each twelve inches high, and, as the pressure is equal, these are drawn over simultaneously. This principle is applied in practice even to one hundred and fifty pipes subjected to the force of the same vacuum, so that those of a whole district are exhausted at the same moment.

In addition to the difference of inclination, there is also a great difference in the quantity of material to be treated, and these different quantities are equally well managed by the same system. In Figure 26, c is the main pipe connected with the vacuum chamber.

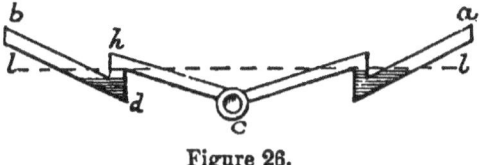

Figure 26.

We will suppose a to be the outlet pipe of a large hotel, and b that of a small cottage in which only two persons are living. The pipe a receives an amount of liquid which will fill the space below the lines l l in an hour. During the remaining twenty-three hours of the day its sewage matter flows on directly toward the central reservoir; but the accumulation in the pipe b is only sufficient during twenty-four hours to fill the vertical part of the pipe between h and d. Of course this matter will lie level in the angle, and will rise but a part of the distance between d and h. When the vacuum is applied, the atmospheric pressure at b bears down upon the small supply and tends to raise it toward h, but at the same time an equal pressure in the pipe a is forcing forward the contents of that pipe and pouring them over the height. The contents of b cannot reach the point h until the quantity in the pipe a is reduced to the same amount, that is, until the whole pipe between d and a and d and b is emptied; then there will stand in the two pipes two columns, each twelve inches high, ready to pass over at the same moment.

This device has enabled Liernur to do away with every faucet or stop-cock in his whole system of pipes, except a single one in the main. By opening this the force of the vacuum is brought to bear equally and instantly upon the house pipes of the whole system, with a quick pneumatic shock whose suddenly applied force is deemed important. It is thus made certain that there can at no point be a useless escape of air, until every one of the pipes has been exhausted of its contents; of course, at the angle, a small quantity will fall back after the air begins to flow over.

The arrangement of house closets is very simple: they are, wherever practicable, for economy's sake placed vertically one over the other on the different floors, in order that they may reach the outflow through the same down-pipe. The closet, as originally made, is a simple funnel of iron or earthen-ware with a bend trap at the bottom, as shown in Figure 27, a pan of enameled iron or whitened earthen-ware being inserted at the top for better appearance. From the highest point of the main pipe, outside of the trap, there rises a ventilating pipe, v, reaching above the top of the house, and this pipe has a branch for the ventilation of the funnel, which it enters near its top, at a point behind the pan. The action of

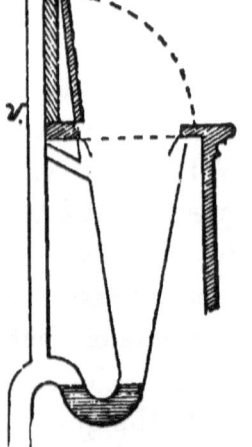

Figure 27.

this branch is to afford an outlet for gases forming in the funnel and to cause a down draught when the lid is opened, so that there may never be an escape of foul air into the room. It is recommended, when practicable, to place these closets next to the outer wall of the house and to supply each with an open window, or in some manner to give a thorough ventilation. The pipes descending from the closets, the service pipes of the different houses, and the mains in the streets (in each district) are all five-inch cast-iron pipes, secured at the joints in the same manner as gas pipes.

So far as the emptying of the closets is concerned, it is thought that the system, as described, is entirely complete and satisfactory. The next problem was to apply it to the solid matters of the kitchen waste pipe. The amount of water flowing from the kitchen, from bath-tubs, etc., is much greater than it would be economical to treat by the pneumatic process, and a separate outflow is provided for them to the same system of sewers that is used for the removal of storm and subsoil waters. Figure 28 shows the arrangement of the kitchen drain apparatus: *a* is a reservoir, say one foot square, furnished four inches below its top with a grate or screen fine enough to prevent the escape

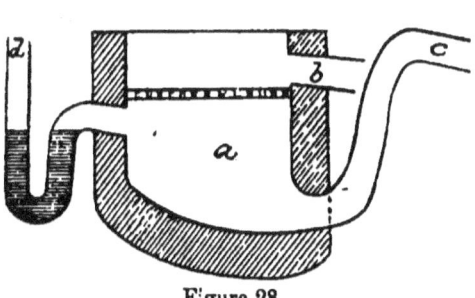

Figure 28.

of any coarse matters which might obstruct the street sewer, or which it is worth while to preserve as manure. The bottom of the reservoir is curved, and is connected with a pneumatic sewer pipe; the outlet *c* takes, immediately, the rise of twelve inches needed to preserve the barometric resistance. The house drain *d* discharges its contents into the reservoir below the screen; it has a bend trap deep enough to give a decided resistance to atmospheric pressure. The flow from the house passes into the reservoir *a*, and its excess of water rises through the screen and flows off at *b*. During the day, more or less solid matter is accumulated below the screen, and when the pneumatic pressure is brought to bear, by opening the main pipe near the vacuum chamber, it is, simultaneously with the closet pipes, emptied of its contents, and at the same time whatever matters have adhered to the bottom of the screen are forcibly blown away by the pressure of air descending through it. In this way, while the chief volume of water or other liquid matters is got rid of at once through the sewers, the more valuable solid material, which would create inconvenience in the sewers, and which has a manurial value, is added to the products of the closets for treatment with them during the subsequent processes of the system.

A locomobile engine having somewhat the appearance of a steam fire-engine, carrying a steam-engine and air-pump, and followed by a tender in the form of an iron tank, to which its air-pump may be attached, is used during the construction of the work, before the different street reservoirs are connected with a main central pumping station. The air-pump is attached to the opening at the top of the street reservoir, from which it exhausts the air, making about a three-quarter vacuum. The cocks in the mains being opened, the house-wash of the district flows into the reservoir, which is then closed, and the air-pump exhausts the tank of the tender. Then this is closed and its supply pipe is connected with the pipe reaching to the bottom of the reservoir, when, the valves being opened and the air being admitted to the top of the street reservoir, the contents of the latter are sucked into the tank, which may be driven away to the point of discharge.

This locomobile serves to demonstrate the practicability of the system, and is an indispensable accompaniment of the earlier steps of construction. But its purpose is only a temporary one, and as fast as may be the street reservoirs are connected with the central station, by pipes which it is often necessary to make larger than five inches owing to the quantity of liquid to be discharged through them. Each central station may answer for a district of say fifty thousand or sixty thousand inhabitants.

At this station a fixed engine and large receiving tanks serve for the numerous street reservoirs the same purpose that these (with the locomobile) originally served for the houses of their separate districts. The tanks at this station have sufficient capacity to receive the contents of the whole set of street reservoirs with which they are connected, and the engine has a sufficient power to maintain the required vacuum in these and in the main pipes. By precisely the process heretofore explained, the contents of the reservoirs are drawn to these tanks, and are made ready for their subsequent treatment.

The receiving tanks at the central station, which may be one or more in number, are large enough to store the contents of all the street reservoirs of the district. They are located in the basement, and each has an indicator by which the engineer can see when it is filled. We will now assume that all of the street reservoirs have been emptied, and that the tanks in the basement are filled. These tanks communicate by suction tubes with a similar tank elevated above the main floor of the building, which has also an indicator showing the level of its contents. This upper tank is exhausted of its air by the air-pump, and the communication between it and the bottom of one of the tanks in the basement being open, it fills itself with the liquid, which is now ready to be treated by the poudrette apparatus. For this purpose it is allowed to flow into a vertical tank, in the bottom of which there are coils of pipe connected with the exhaust pipe of the steam-engine.

The steam, on its escape from the exhaust valve, passes through a coil in a superheating chamber where the products of combustion on their way to the chimney, flowing around the coil, give the steam an additional heat. This reheated steam passing through the coils in the evaporating tank produces a furious ebulition and a rapid evaporation of the water of its contents. The condensation at the next stage of the process of the vapors thus formed, tends to produce a partial vacuum above the boiling liquid, so that this rapid evaporation may even take place at a temperature below that of boiling water. The condenser into which these vapors pass is a copper drum, the temperature of which they raise probably to two hundred degrees Fahrenheit. This drum revolves slowly, its lower part passing through the semi-desiccated, pappy liquid drawn from the evaporator first described. As it makes its slow revolution it carries up a film of the pap, which the heat within renders perfectly dry, so that near the end of the rotation it may be scraped off by a stationary knife, and fall into a receiver below in a desiccated state, ready to be packed in bags or barrels for agricultural use.

This desiccated poudrette contains all or nearly all of the organic refuse of the household, not only the contents of the closets, but the particles of unused food, grease, and other solid constituents of the kitchen waste. The chief difference in condition between it and guano, or the manufactured poudrette of commerce, is that the matters it contains have had no opportunity to pass into a state of decomposition. Ordinarily, within thirty-six hours from the time of their production in the house they have all been transported to the central station without exposure to the air, desiccated, and packed away. As during the evaporating process a small quantity of sulphuric acid is added to the liquid, any ammonia produced by incipient fermentation is rendered non-volatile.

Concerning the value of this Liernur poudrette I have no other evidence than the following account of Professor Voelcker's analysis given in Mr. Adam Scott's description of the system, in the "Sanitary Record" of November 21, 1874.

An analysis by Professor Voelcker, chemist of the Royal Agricultural Society, dated August 15, 1874, of a sample submitted to him by Sir Philip Rose, Bart., showed it to contain: —

Moisture	8.64
Organic matter [1]	62.96
Oxide of iron and alumina	3.29
Phosphoric acid	1.76
Lime	0.86
Chlorine	6.22
Sulphuric acid	6.02
Alkaline salts	8.20
Silica	2.05
	100.00

So far as I have been able to learn there has been no sufficient practical test made of the value of this poudrette, but when we consider the substances from which it is produced, it seems impossible that it should not have a great value, and Liernur and his advocates bring ample theoretical evidence in support of its claims. If it is true that the waste of the constituents of food which characterizes the do-

[1] Containing nitrogen 9.35, equal to ammonia 11.35.

mestic habits of all our towns is leading to the ultimate impoverishment of our fields, we can hardly regard with too much interest any process that promises to restore so nearly the entire amount of their products consumed and squandered in our households.

Mr. Scott, in the article referred to, thus describes the practical working of the system:—

" The air-pump engine is set in motion, and maintains during the day a three-quarter vacuum in certain central reservoirs, placed below the floor of the building, and at the same time in the central pipes. Workmen perambulate the town, visiting each tank once a day. To drain the houses commanded by one tank, they alternately open the connecting cock of the central pipe and the stop-cock of any main pipe ; the first to obtain a vacuum in the tank, the second to utilize this by emptying the closet-pipes connected with that particular main. After all the mains of the tanks in question have been operated upon, and their contents collected in the tank, the workman turns the discharging cock to send the whole mass to the central builing for immediate conversion into poudrette. He then proceeds to the next tank, there to repeat the operation."

One of the minor objections anticipated by its inventor to the general introduction of this system is to be found in the fact that an influential class in every community where the water system has been introduced may object to any less fastidious substitute for the water-closet. To meet this objection there has been devised an apparatus, in which water is used, which seems completely to compass the requirements, but the practical need for its use is too slight for it to be considered as an essential part of the system. And indeed it is better that at every step of the process there should be as little extraneous water as practicable thrown into the pipes. The natural product of liquid matters in every household is sufficient to insure the proper pneumatic action, and all additions beyond this create an increased demand for fuel for the final dessication.

It is thought by the advocates of the pneumatic sewerage that all other systems thus far tried, in addition to their danger to the public health, are necessarily and always very expensive, there being no offset in the way of profit that can possibly lessen the taxable charges required for their construction and operation. It is believed also that these taxable charges are an excuse for the raising of rents, and consequently for the crowding of the working classes into smaller and less commodious and less healthful quarters than they might have were the town free from the necessity for making this excessive yearly outlay.

It is no doubt too early in the history of pneumatic sewerage for figures based on actual experience to be adduced in support of its economy, but the published estimates, which so far as one can judge are based entirely on similar uses of steam, cost of laying pipes, etc., and which are apparently reliable and correct, show that so far from being a source of expense, the fæcal matters of the town may constitute a reliable source of income. Such estimates have too often to be modified, in the light of subsequent experience in actual practice, to be relied upon with great confidence, but there seems to be a sufficient margin to cover any unforeseen contingencies and still to leave an important amount to be credited against the costs of working.

It is stated that the cost of the work in Amsterdam, including royalties, engineering, plant, machinery, and the necessary changes in houses, was not quite £2 10s. per inhabitant. To be on the safe side, Mr. Scott estimates that the cost in an English town would be £4 per inhabitant, and he applies his calculation to a town area of 250 acres, with a population (75 per acre) of 18,750, placing the total cost of the works at £75,000. So far as the Liernur system alone is concerned, without refering to the storm-water sewerage, the cost would be, *pro rata*, the same for a small town as for a large one, provided the population is of the same density.

" Using the figures and proportions given by Captain Liernur, the following would be the estimate of working expenses per day : —

Coal, — Power of air-pump engine required, 80 indicated horsepower. Consumes, at 5 lbs. per horse-power per hour, in twelve hours, 4,800 lbs. coal. Of the caloric due to this there is converted into work eight per cent., or caloric due to 384 lbs., leaving the calorics of 4,800—384=4,416 lbs. on hand for evaporating purposes. There is, however, to evaporate 54

LIERNUR'S PNEUMATIC SYSTEM OF SEWERAGE. 303

ounces per day for 18,750 persons, making 63,281 lbs. water, requiring with drying apparatus à *double effet*, 63,281÷12= 5,273 lbs. of coal, for which there is left the above 4,416. There is hence wanted 5,273—4,416=857 lbs. additionally to the 4,800 lbs. of the air-pump engine, making in all 4,800+857 =5,657, or say 2½ tons of coal per day, which, at 25s. per ton

	£	s.	d.
gives	3	2	6
Oil	0	4	0
One machinist and eleven laborers	2	0	0
Administration, repairs, and sundries	0	13	6
	6	0	0

Making per year, £6×365 . . . . . . 2,190 0 0
To this would have to be added, —
For interest on capital of £75,000 borrowed from local board, including redemption, at four per cent. per annum £3,000
For renewal fund of machinery, at eight per cent. on £3,000 240
————3,240 0 0

Total expenses 5,430 0 0

" The income would be, however, the poudrette manure of 18,750 persons, which, at 10s. per head, gives annually the sum of £9,375, leaving, after deducting above expenses, nearly £4,000 annually as clear profit, after paying every charge."

This calculation is based on an estimate of ninety per cent. of water and ten per cent. of solid matter in the liquid as it is received at the central station. By an application of the same data to liquid containing ninety-five per cent. of water, the cost of evaporation with coal at twenty-five shillings per ton would be £1,081 in addition, which would reduce the net profit from £3,940 to £2,869. It is to be observed that with us his data would have to be materially changed, the cost of coal and labor being much greater, interest being at least six per cent. instead of four per cent., and the agricultural value of the product being certainly no larger.

What has been thus far given covers my knowledge of the Liernur system as derived from the various publications concerning it. It seemed worthy of further investigation, and I devoted some time to its study during a recent visit to Europe.

At Captain Liernur's office, in Frankfort-on-the-Main, I was shown the working drawings of every part of the system, and had all its details clearly explained by its very intelligent inventor, who to a thorough familiarity with modern sanitary engineering adds the most unbounded and enthusiastic belief in the merits of his own invention. I learned that steps are now being taken for an important trial in the city of St. Petersburg, at the hands of a company, who, upon its success being demonstrated, hope for a concession for the sewerage of the whole town. The conditions there existing are the same as in other places where actual trials have been made, save that the intense cold and the consequent necessity for placing the apparatus deep below the surface of the ground must increase the cost of construction, and, so far as house-pipes are concerned, may present many difficulties to be overcome. The use of the system at military barracks in Austria and Hungary was described as having been success ful and profitable, but I was directed, for an ocular demonstration of pueumatic sewerage in actual operation, to visit Amsterdam and Leyden, in Holland, where the earliest trials were made, and Dortrecht, where the whole invention in its entirety is being adopted.

At Dortrecht, Liernur's partner, Mr. De Bruyn Kops, is constructing works for a large part of the town, to be subsequently extended over the whole. The central station was nearly finished, and contained a thirty-five horse-power steam-engine, and an air-pump suited to its capacity; basement tanks capable of holding two days' product of the whole town; an elevated tank through which to transfer the liquid to the poudrette apparatus; and this apparatus itself, which was complete and had been in use. The superheating effect of the escaping products of combustion had been found insufficient, and a separate furnace with a small fire had been provided to raise the heat of the steam to the required point. The attempt to manufacture poudrette had not been entirely successful, that is, the product was rather moist and pasty than dry, and some modifications were being made in the machinery which rendered it impossible for the station to be at work during my visit. Pending these repairs the street reservoirs were being emptied by the locomobile, but as I was to see this in operation in Amsterdam, it was not thought worth while to bring it out. From the station we visited the poorest quarter of the town, in which the pipes had been laid, passing through a district that still depended for its cleansing upon a sluggish canal, — a canal of the most offensive description, its surface constantly bubbling with the gases of the decomposing filth it contained. Similar canals had been filled up in front of the houses connected with the pneumatic system, and this of itself should be a sufficient improvement to satisfy the Dortrecht authorities with their outlay. We visited closets in houses and in yards, and so far as I could judge from the manner of those who exhibited them, these were perfectly satisfactory in their operation. Equally unobjectionable closets in the houses of people of a corresponding class I have never before seen, and my general impression of the condition of the work in this town was that it may be in a fair way to prove all that its inventor claims for it, except possibly in the manufacture and value of the poudrette.

The next day we went to Amsterdam, where (and at Leyden) the first experiments with the system were made. It is now in universal use in nine considerable sections of the town, and is being gradually extended. The poudrette apparatus is not in use there; indeed, the only set thus far put up is the one now being experimented with at Dortrecht. At all the stations in Amsterdam the liquid is run into barrels and transported to the country by canal-boats, being sold, thus far, for a nominal sum, very much less than would be its value here.

At the first station which we visited the engine was out of order, and we could see nothing; but at the second station it was demonstrated in my presence that the working of the air-pump and its effect on the street reservoirs of its district are entirely satisfactory. The liquid was transferred from house pipes to several street reservoirs, from these to the basement tanks at the station, and from these to the elevated tank from which the barrels are supplied, with certainty and regularity. In one case it was necessary to carry a main pipe, by a siphon, under a canal, and the transferring of the liquid through this was entirely successful. Indeed, if the object were only to transport in a quiet, inoffensive, and entirely hidden manner the products of private houses to a depot whence they can be inoffensively shipped to the country, my investigation seemed to prove clearly that entire success had been attained.

I hoped before leaving Holland to be able to see the Dortrecht poudrette works in successful operation; but a further trial, although it showed a great improvement, left something still to be desired, and the apparatus was not in satisfactory working at the time of my leaving the country.

In Amsterdam we visited a great number of houses of all classes, — a large children's hospital, private houses of the best class, tenement houses occupied by working people, an old ladies' home, and in one case a nest of sailor boarding-houses, which were said to be the worst in the whole town. This examination was of course made under the guidance of one who was interested in the success of the system, and it is possible that, had I been conducted by one opposed to it (and there are such), I might have been shown instances of failure. As it was, I can only say that under all the circumstances and conditions, both where the greatest attention was given to cleanliness and where the greatest neglect seemed to prevail, I found the condition of affairs in all cases good, and among the lower classes infinitely better than would be found in similar establishments in London or in New York, where the water system and the common vault prevail, though to the eye a well-kept water-closet is preferable.

Subsequently I took occasion to talk with several gentlemen of intelligence in Holland about the success and the prospects of the system. Of these, none were opposed to it, and some favored it very strongly. Mr. Van der Poll, the Dikegraaf of the Haarlem Lake Polder, who is an engineer of high standing and of sound judgment, gave it as his opinion that it must inevitably come into universal use in all the towns of Holland, although he was not prepared to say that it is better than water sewerage for places where a good and suitably located outfall can be had. Another friend was glad to get my opinion, for the reason that so much passion had been shown in all discussions of the subject in Amsterdam that it was impossible for disinterested persons to weigh the evidence for or against it. It was stated that there had been very serious opposition, and that the early introduction and working had been embarrassed by the fiercest opposition of the chief official who was directed with its execution, but that in spite of this, and of all the drawbacks attendant upon the education of the people in a new process, and all the mistakes inseparable from the practical development of a new invention, it had steadily made its way in popular favor, and had especially won the approval of the city officials, under whose direction it is now carried on. (An official told me this.) In one instance a large speculator in real estate, one who buys blocks of ground and builds houses for sale, had been originally a very strong opponent, protesting most earnestly against the introduction of the system in districts in which he was interested. He is said now to petition for its introduction in each new district in which he buys property.

These statements are made with the reiterated qualification that my investigation was made under the guidance of one who is pecuniarily interested in the invention, and who had it in his power to mislead me, but who, I am glad to say, impressed me as a frank and fair-minded gentleman, who made no attempt to conceal defects, or to bias my judgment. Since my return I have learned that Dr. Folsom, Secretary of the Massachusetts Board of Health, found his inspection of the working of the system in Amsterdam very unsatisfactory.

The question that naturally suggests itself is whether Liernur's pneumatics are to solve the whole sewerage problem. It would no doubt be safe to answer this, at once, in the negative, but it should be a negative with many qualifications. The whole problem is now so entirely unsolved, and is so embarrassed with intricacies and difficulties at every turn; it is of such vital consequence when regarded from the point of view of the public health; and it appeals so directly to the strongest interest of every householder, that no one interested in the subject can fail to give very careful attention to any suggestion of relief which promises so much as Liernur's does promise, and which is in all its details so complete and so well-balanced, and is apparently so successful in each department of its mechanical action.

On the other hand, we have been so long relying on the system of water carriage, and we have so long ascribed to it every advantage, only to find it riddled and honey-combed with faults, as time has brought us better acquainted with it; and a large class has placed such implicit confidence in the dry-earth system, only to find it almost impossible of introduction in an average community, that no one who has been long interested in the general question can be expected to glow with enthusiasm over any new process that may be brought to notice. Liernur has struck out a new path, but it is a new path in an old field, in which we have learned to look out for pitfalls and ambushes at every step. We may well hope (and I unreservedly believe) that there is much in his invention that is of intrinsic value, and that it will perhaps accomplish all that we have so long sought. At the same time its success is certainly not to be achieved through a blind enthusiasm, ready to accept it as the final cure of the great and universal disease in our domestic economies against which it proposes to contend.

While, therefore, it is to-day unquestionably the most interesting new fact in sanitary engineering, and is worthy of the most careful experiment and even the most expensive investigation at the hands of local governments, the investigation and the experiment should be made with a clear understanding that the time given to them and the money spent upon them may bring but little return. The difficulties we are contending with are so grave, and the dangers to life and health and usefulness are so threatening, that we may well afford to tax ourselves as largely as may be necessary in order to demonstrate whether this new process, for which so much is claimed and which has so many firm adherents among those who have been living under its daily operation for some years, is or is not to open the door for our escape. Much that has hitherto been written about it has been of that enthusiastic and confident character that made its success appear at first blush a foregone conclusion. It seems to be better that, however great our individual confidence may be, — and I repeat that my own is very great, — we should undertake this trial resolutely and determinedly, but should at the same time be quite prepared for entire or partial failure.

The more ardent advocates of the system lay great stress upon its economical features, and seem to depend very much upon the prospect of profit for the reënforcement of their arguments. Let us rather take the wiser course of throwing the questions of profit and economy entirely into the background, where they belong. This is a subject that reaches much farther than any pecuniary interest, and it is one whose pecuniary interest centres much more in the lengthened life and full, healthful efficiency of our populations than in any question of the cost of constructing works, or of proceeds from the sale of manure. If it is found that with our price of machinery, labor, fuel, interest, and manure we can sell the product of Liernur's poudrette apparatus or the liquid drawn from Liernur's vacuum tanks at a price that will give a profit, or even will help materially to defray the expense of the system, it will be so much gained; but our people are quite prepared to take such a view of the sanitary question as makes all this far less than secondary. If the elements of fertility can be saved for return to our fields, and so continue and increase our prosperity, the benefit resulting will be immeasurable; but this benefit is, to the common understanding, too vague and theoretical to have much influence on the minds of the average denizens of towns.

Any prudent community, interested in the reformation of its present health-destroying process, will naturally and properly set aside all considerations of this character, and make their investigations of Liernur's pneumatic sewerage, or of any other system that may promise them relief, with an almost sole view to the completeness of its sanitary advantages, and to its practicability from a mechanical and commercial point of view.

All that it is safe to say about the system now, in

its relation to our own condition, is that it is, as regarded in the light of what we know about the water system and the dry-earth system, sufficiently promising to justify the most energetic investigation. So far as I know, its opponents have adduced nothing against it that may not be remedied by practicable mechanical improvements, and its advocates, who are many, speak of its advantages with a confidence that, often at least, has grown from favorable experience of its practical working.

CHAPTER X.

THE DISPOSAL OF SEWAGE.

THE problem of sewage disposal is always serious, and it becomes more and more so, as population increases, and as sewered towns multiply. How to get rid of the offscourings of any community in such a way that there shall be no return of offensive and dangerous odors; that there shall be no accumulation of foul matters inconveniently near; and that there shall be no tainting of the source of the water supply of other towns, is a question which taxes, and often overtaxes the ingenuity of the engineer, and the paying capacity of the finance department.

The limits of this book would not suffice for an adequate description of the varied experience of English towns in this matter. After years of trial, and millions of expenditure, the authorities on the subject are widely at variance as to what may best be done, and the more prudent of those who have given thought to the matter seem as far as ever from accepting any result yet accomplished as satisfactory.

Many schemes are urged by enthusiastic advocates, as offering sovereign remedies for the evil. All of these attach more or less weight to the value of the material under consideration for fertilizing purposes, and those are not few who have hoped to derive from the use of sewer water large profits in return, from the adoption of their various devices.

I can only say here, that none of these schemes have so far achieved the success claimed for them, as to gain the confidence of the engineering world at large. The facts remain that the material in question is extremely troublesome; that it must be got rid of in some way at all hazards; that it must not be allowed to injure the health of those producing it, or of other communities, and, worst of all, that its manurial value seems to be less than is needed for its profitable use (unless in a few special cases) under the plans thus far devised. Without going at all into the question of the disposal of the sewage of large towns, farther than it has already been considered in earlier chapters, I propose merely to call attention to one or two devices which seem to afford relief in the case of small villages, and especially of large or small private establishments.

IRRIGATION.

As a general principle it may be stated that in sewage irrigation the amount of land appropriated should not be less than one acre to one hundred and fifty of population, and should lie not more than a mile from the town. The same land should not receive sewage two days in succession, and each area should have occasional periods of rest for a whole growing season.

If the land is of a very retentive character, even if well underdrained, it would be better to allow one acre to one hundred of population.

This applies to the ordinary irrigation of agricultural fields, by surface flow, but Mr. J. Bailey Denton, who is an old, and very accomplished drainage engineer, adopted a new system in treating the sewage of Merthyr-Tydvil, in Wales, which has there had a success that seems fully to justify its repetition wherever a suitable soil, and a sufficiently mild climate admit of it. In this case the purpose was not especially to turn the sewage to profit, but to purify it of its organic matters, so that it could be run into a stream without polluting its waters. Mr. Denton calls this system "Intermittent Downward Filtration." A gravelly soil is thoroughly underdrained at a depth of six feet, and is divided into two separate plots to which the sewage is applied alternately. After a certain amount has passed through one field the supply is turned on to the second, and the first is allowed to become thoroughly aerated, and so cleansed by the oxidation of the organic matters that it has taken up, as to be ready again to serve its purpose as a filter.

The latest report from these irrigating fields, which have now been several years in use, comes from Dr. Dyke, the health officer of Merthyr-Tydvil, whose last annual report bears testimony in favor of the system of intermittent downward filtration. Speaking of a plat of twenty acres laid out to receive sewage on Mr. Bailey Denton's plan, Mr.

Dyke declares that the system is cleanly, odorless, and perfect in all its details. As a part of his duty, he has from time to time examined the water flowing from the outlet, and has satisfied himself that there has been no perceptible increase in the amount of its organic matter. In July, 1872, it was proved by the investigation made by Dr. Frankland to contain only one part of organic matter in 200,000 parts of water; and in 1874 it was found still just as free from impurity, showing that no saturation of the soil with filth took place, the growth of vegetables on the surface, and the aeration effected by drainage doing the work of purification effectually. In another part of the report, which treats of the classes of disorders met with in the district, Dr. Dyke recalls certain prophecies, made when the sewage-farm system was introduced, that the vegetables grown under its means would act prejudicially on the public health. So far from this being the case at Merthyr, there are hundreds of young children brought up on the milk of stall kept cows, fed partially on grass grown on sewage-irrigated meadows; yet the mortality among them from diarrhœa and similar complaints is exceptionally low, and all medical men in the district agree that this type of disease among persons of all ages has diminished of late years, for the plain reason, no doubt, that whilst sewage-irrigated vegetables have been introduced, a good water-supply and a thorough system of drainage have been introduced with them.[1]

[1] *Sanitary Record*, October 30, 1875.

ARTIFICIAL PURIFICATION.

The following account of the experience at Coventry (England), illustrates very well one of the more successful cleansing processes.

The corporation of Coventry, having been ordered by the Court of Chancery to discontinue the pollution of the river Sherburne, contracted with The General Sewage Company for the treatment of their sewage. The works were completed in April, 1874, since which time they have been constantly working in the most satisfactory manner. An average of two million gallons are treated daily. "The sewage, strained and freed from its grosser contents, passes through a block of buildings where it receives continuously a charge of sulphate of alumina in solution, and is thoroughly mixed therewith. These buildings are well provided with steam-engines, boilers, and mixers, as well as ample machinery of a first-class character, adapted not only for the admixture of the chemicals with the sewage, but also for the manufacture of the chemicals employed. The effluent water is in a sufficient state of purity to enter most rivers or the sea. It is, however, in this condition, subjected to a process of filtration, and for that purpose is conveyed to a filter-bed properly drained and prepared. It has been found that a comparatively small quantity of land (four acres and a half) would suffice to filter the effluent water from the sewage of Coventry. Nine acres have, however, been prepared, for the purpose of always having another filter-bed ready when the one last in use requires rest or repair. The effluent water, as it passes into the river Sherburne from the three large mains of the filter, at the rate of about 80,000 gallons per hour, is clear and bright, and not only so, but of a high standard of purity, as is shown by the analysis of it made by Dr. Voelcker, who says: ' The water has no perceptible smell, and is almost free from color; it contains but little organic (albuminoid) ammonia, and not much more than half a grain of saline ammonia per gallon. And, further, that the nitrogenous organic constituents of raw sewage appear to have become oxidized and changed into nitrates to a very large extent.' The sewage of Coventry, at noon, has been found to contain as much as 5.85 parts of ammonia in 100,000 parts."

The sludge or solid precipitated matter amounting to about twenty tons per day is dried and sold for manure. The process requires chemicals to the value of £2 12s. 10d. (say about $13), for the purification of one million gallons. The whole sewage operation at Coventry cost per annum about 5d. or 6d. per head of population.

THE DENTON AND FIELD STORAGE TANK.

In order to secure an intermittent flow for the general application of his downward filtration system, Mr. Bailey Denton has, in connection with Mr. Field, applied the principle of the Flush Tank to the use in agricultural irrigation of the sewage of small communities, — where the constant stream is too slight to secure the flooding of a sufficient area for an economical use of the sewage, and for intermittent application to successive fields.

At the hamlet of Eastwick near Leatherhead, in Surry, this system has been in operation for the past three or four years and as this is the oldest experiment with a new and apparently very valuable device, it seems worth while to reproduce the description of it from Dr. Simon's report of 1874 (Netten Radcliffe's paper).

" Eastwick is a hamlet of fifteen houses, including the mansion of the proprietor and the farm homestead ; and it has a population of about one hundred and forty-five. In devising a system of excrement and slop disposal for the place, any general plan of water sewerage had to be set aside, the water derived from wells being variable in quantity, and at no time too abundant for ordinary domestic use, irrespective of water-closets. The common privy was retained for the cottages, but the privy-pit was converted into a water-tight receptacle beneath the floor of the closet, and the cottagers were instructed to throw into it above the excrement, the refuse ashes, and to remove the contents of the pit monthly for use in their gardens. Four water-closets exist and five earth-closets for the use of the mansion and its precincts; and one water closet and three earth-closets for the use of the farm homestead. To provide for the liquid house refuse of the hamlet, and for the drainage of the farm buildings, the scheme of sewerage was carried out by Mr. Bailey Denton, which is shown in the accompanying plan, and which has an outlet in a meter tank, of which the plan and section are given in the following Figures."

"The tank is in two compartments to admit of cleansing without entire disuse. It has a capacity of five hundred gallons, and it fills and discharges in ordinary dry weather three times in two days. The several discharges are directed successively in different portions of a plot of ground prepared for the purpose, and which measuring three roods three perches, serves ordinarily for the effective and profitable utilization of the whole liquid refuse of the several cottages, the mansion, and the farmstead. The drainage of the latter includes the flow from cattle sheds and stables, in which from fifteen to twenty animals are always present, and about thirty head of horned cattle, and thirty horses at intervals. The drainage of a large piggery also passes to the tanks.

"Luxuriant crops have been grown upon the irrigated land, last year's crop consisting of the thousand-headed cabbage. Of this crop, Mr. Hutchinson, the steward of the estate, says: 'Besides thriving so well upon the sewage, it is an excellent food for milk cows, being less strong in taste than the drumhead and not having any but a good effect upon the milk. The thousand-head can also be used as human food. I estimate the value of the crops obtained at £25 per annum, or at the rate of £32 10s. per acre; and the outlay in attendance

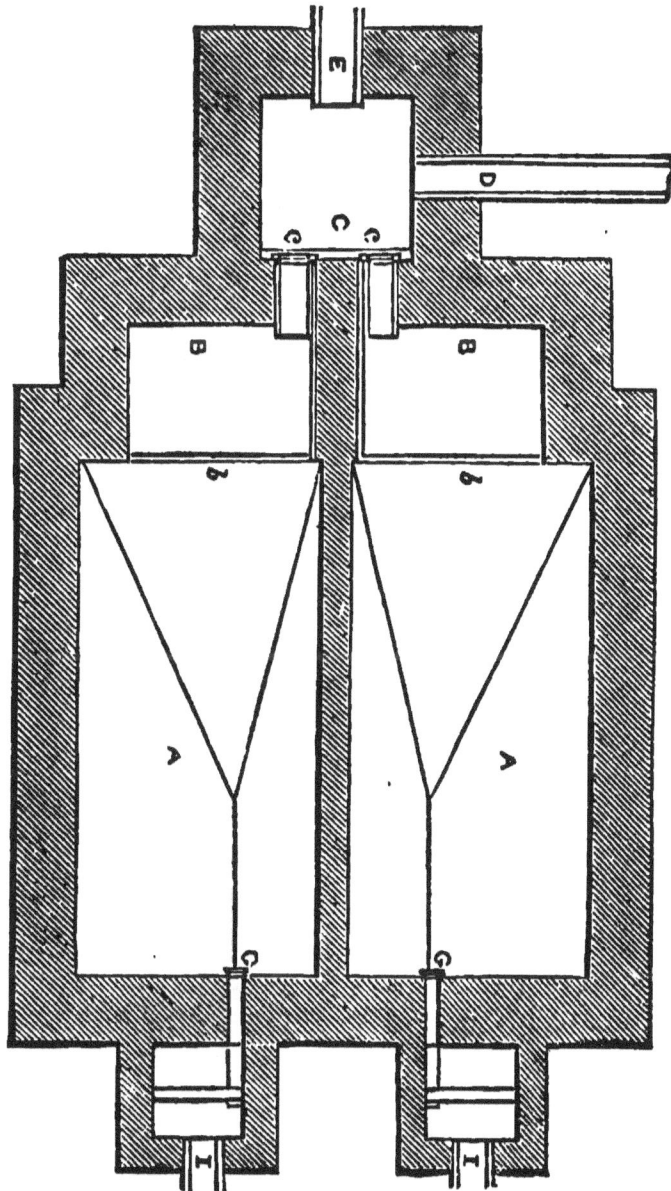

Figure 29. — Denton and Field's Sewage Meter, Plan.

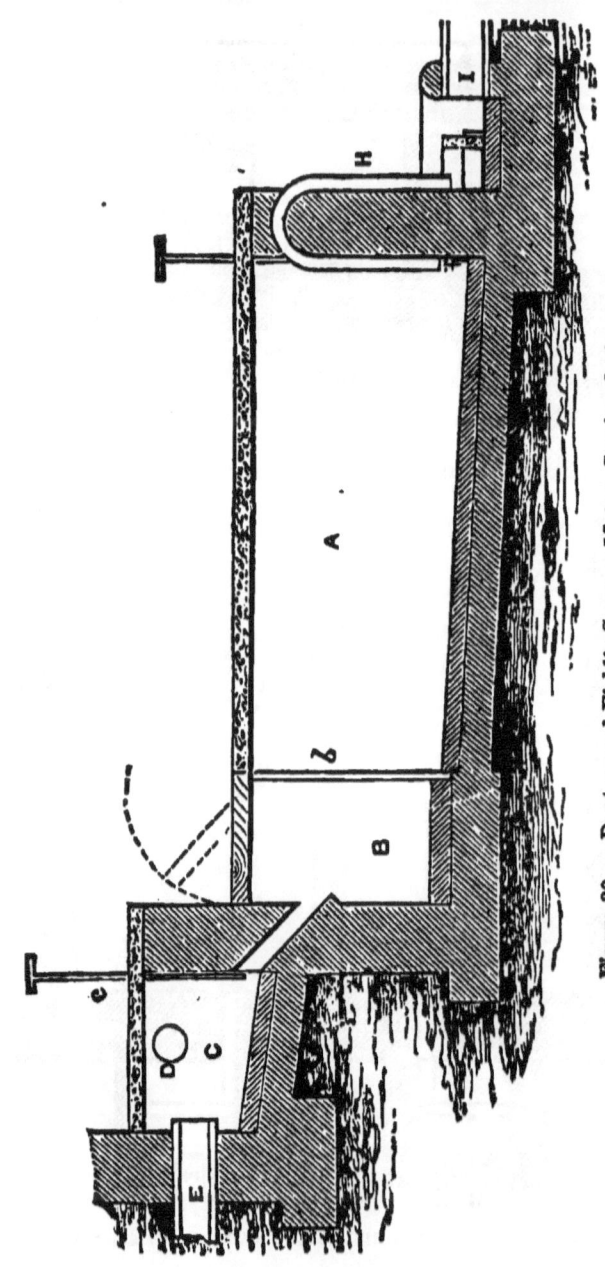

Figure 30.—Denton and Field's Sewage Meter, Sectional view.

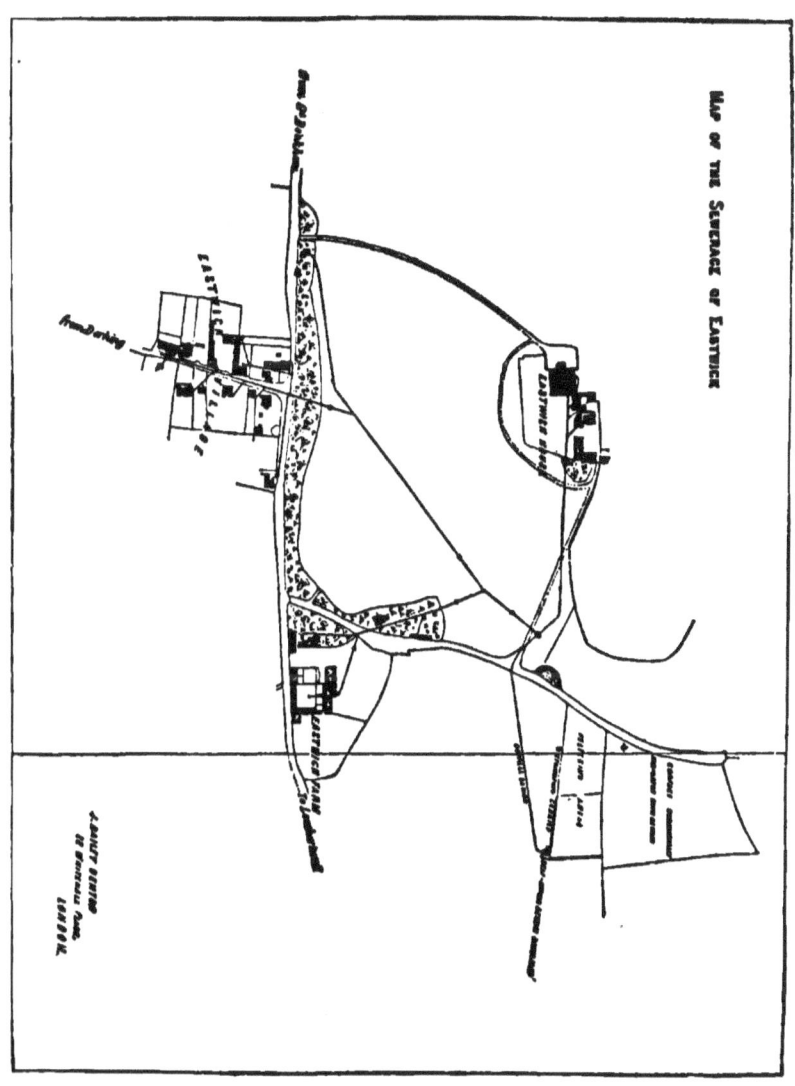

Figure 31.

upon the land and the regulator ("meter") I put down at £7 16s.'

"Mr. Bailey Denton, to whom I am also indebted for the plan, has courteously sent me the following statement of the cost of the works above described, including the 'meter' and the preparation of the land, and he remarks upon this statement that, 'the yearly return, after deducting the cost of attendance upon the sewaged land and regulator cannot be less than £17 per annum, so that already a return of about five per cent. on the outlay is gained, while there is every prospect of increasing that return as the quantity of sewage dealt with becomes greater and its treatment becomes better understood.

EASTWICK SEWERAGE.

	£	s.	d.
To payment for labor	179	4	0
pipes	103	7	2
stone, lime, cement, and sand	12	14	10
iron and lead work	20	5	1
carriage of materials	1	9	1
Traveling and incidental expenses	3	12	0
	320	11	4

"In regard to abatement of slop nuisance, and I may add also largely of farm nuisance, among a rural community, the arrangements at Eastwick are the most complete and satisfactory I have yet seen. Notwithstanding the contiguity of the irrigated land to the mansion, no nuisance is experienced from it, whereas previous to the present arrangements, when the slops of the mansion and cottages found their way into neighboring ditches and decomposed there, considerable nuisance had existed. With some structural alterations in the privies (the principles of which are stated in their proper place in this report), and such needed supervision as will now be obtained from the sanitary authority appointed under the Public Health Act, 1872, the arrangements at Eastwick may be regarded as a pattern to be followed by villages and small towns similarly circumstanced.

"From what has already been said it may be inferred that the 'Automatic Sewage Meter' admits of wide application in removing the difficulties which often beset the disposal of the sewage of communities larger than Eastwick. It simplifies the whole question of dealing with the sewage of small towns, villages, isolated institutions, and mansions, while securing the most efficient application of the sewage to land, both for purification and utilization, with the least expenditure of labor."

J. A. Davenport, in his paper on Village Sanitation and Rural Drainage and Habitations, says with reference to the disposal of slop water through open drain pipes laid near the surface : —

"I have in my district a block of eight or ten houses (belonging to Mr. Owen Lant of Nantwick) that have been drained on this system for two years past, and all is working well up to the present time, with no signs of choking. Most of this class of work is roughly done, and under conditions rendering a good result a matter of some doubt, still so far it is as satisfactory as might be expected. A considerable number of houses in this district are drained upon this system, but I mention Mr. Lant's houses as being the first that were dealt with and which have therefore stood the longest test. The difficulties in persuading and arranging as to this particular system have been great, for ordinary people do not quite comprehend it. Many heads have been shaken over it, and some of the attempts at carrying it out would be amusing to relate. The only principle that I have considered it safe to act upon is to deal with all foul liquids by the soil; get them properly on to it or through it as quickly as possible, and in suggesting any rural drainage I have always kept this end in view. The system sketched out, or some little variation of it, which will frequently be necessary, looking at the varying conditions under which such work has to be carried out, will, I venture to think, generally in the country be found to furnish a fair solution of the difficulty. I I have sub-irrigating drains in gardens in my district, where the soil is heavy, and they are at present working well. Probably for villages, etc., no better method of dealing with their sewage could be adopted than that provided by the automatic sewage meter tank. It has been mentioned before, that the irregular flow of sewage on to land (at times a mere dribble, and at other times flowing more copiously) has been a difficulty in its successful application. The 'Automatic Sewage Tank' meets this difficulty. By this means the flow of sewage from the tank may be regulated, it being not at all dependent upon the flow into it."

Mr. Davenport thus describes the Eastwick experiment : —

" From London to Leatherhead by rail, — a delightful walk thence brings one to the little village of Eastwick, the drainage of which is dealt with on this system ; the tank here holds five hundred gallons, discharging itself in dry weather three times in two days, and receives the slops, etc., from thirteen houses, including farm and mansion, representing a population of about one hundred and forty-five. The sewage is utilized upon a trifle more than three fourths of an acre of land. The cost of the works, some £320 may look, in the first instance large, but it is calculated that a return of £17 or something like five per cent. upon the capital, is obtained upon the sewaged land. I was much pleased with my visit here. There was an absence of nuisance about houses, good drainage, and certainly the best method of ultimately disposing of village sewage that I had seen. Before adopting any system of dealing with small quantities of sewage, I would advise any one interested to go and see for themselves the drainage arrangements at the little village of Eastwick."

My own experiment with this system of irrigation by subsoil pipes, which has for the past six or seven years been eminently satisfactory, was instigated by the following which was extracted from an early advertising circular of Moule's Patent Earth-Closet Company in London.

THE DISPOSAL OF SEWAGE.

"HOUSE SLOPS, ETC.

"Where there is a garden, the house-slops and sink-water may, in most cases, be made of great value, and removed from the house without the least annoyance. The only requirement is that there shall be a gradual incline from the house to the garden. Let all the slops fall into a trapped sink, the drain from which to the garden should be of glazed socket pipes, well jointed, and emptying itself into a small tank, eighteen inches deep, about one foot wide, and of such length as may be necessary. The surplus rain-water from the roof may also enter this. Out of this tank, lay three inch common drain-pipes, eight feet apart, and twelve inches below the surface. Lay mortar at top and bottom of the joint, leaving the sides open. If these pipes are extended to a considerable length, small tanks, about one foot square and eighteen inches deep, must be sunk at about every twenty or forty feet, to allow for subsidence. These can easily be emptied as often as required; and the deposit may be either mixed with dry earth or be dug in at once as a manure. The liquid oozes into the cultivated soil; and the result is something fabulous. This simple plan will effectually deal with the slops; there is no smell, no possibility of any foul gas to poison the atmosphere, and with this, and the produce of the earth-closet, any ground may be productive and profitable.

"The two following facts will illustrate the value of this system of dealing with house-slops, etc.

"On a wall fifty-five feet in length and sixteen feet high a vine grows. A three-inch pipe runs parallel with this at a distance of six feet from it for the entire length; the slops flow through this pipe as above described. On this vine, year after year, had been grown four hundred well-ripened bunches of grapes, some of the bunches weighing three quarters of a pound. During a period of four years, for a certain purpose, the supply was cut off. To the surprise of the gardener, scarcely any grapes during those years appeared; but afterwards the supply was restored, and the consequence was an abundant crop; the wood grew fully sixteen feet, of good size and well ripened.

"The other case was as follows: —

"Pipes were laid below two square yards of earth, twelve inches beneath the surface, which were fed with the slops through an upright pipe, about one large watering-potful daily. In the month of November, three roots of Tartarian oats were planted in this piece of ground. The stalks attained one inch and a quarter in circumference; the leaves measured an inch across.

"Several of the ears were twenty-six inches long, and when the crop was gathered eight hundred grains were rubbed out of one ear. The whole weight of corn from those plants was three quarters of a pound. Twelve of these grains were put into the same piece of ground the following year: from these was grown one pound and three quarters of seed. In fact, in a garden of twenty perches, by the use of both solid and liquid manure from one house, three crops were grown in the year, the value of which at market price would be twenty pounds.

"In a garden in which this plan has been adopted for eight or ten years, the pipes were recently taken up in order to see how far they might have been filled with the mud of subsidence. After so long use, very little subsidence was found, and none to obstruct the working of the system, excepting where, in one or two places, the bad laying of the pipes caused some obstruction. There was nothing which might not at any time be remedied in half an hour.

"It will be easily seen that this mode of removing sink-water and slops can be applied to towns or districts of towns. Whilst the application of liquid sewage, in the ordinary sense of that expression, to the purposes of irrigation will be generally impossible, either from the want of proper land or proper fall, or the extravagant cost of pumping, or the difficulty of irrigating during frost or during harvest, this small portion of the refuse-matter of towns, rendered more easy of distribution by the admixture of rain-water, can be pumped to any height, even to land above the town, at all seasons and under all circumstances. During the hard frost of 1867, the sub-irrigation in the garden above mentioned has continued without the slightest interruption."

For how large a community the Denton and Field Storage Tank might be made useful we have as yet no experience to determine. All that can be said for it at the present time is, that so far as private houses, factories, hotels, asylums, and small communities are concerned, it offers a means for the complete solution of the slop question in a manner that will give at least some return for the outlay, for in every case where this underground irrigation is used, there is quite sure to be a greater or less increase of fertility.

NOTE, 2d edition. — Attention is called to the account given in my *Village Improvements and Farm Villages* (Boston: Houghton, Osgood & Co.), of the later application of the system of sub-surface irrigation, especially as applied to the sewerage of the village of Lenox, Mass.

This system is now a demonstrated success, so much so that the Massachusetts Board of Health, in a circular issued in April, 1879, speaks of it as "the best" means of disposal where there are no public sewers, and where water-closets are used.

CHAPTER XI.

THE DRAINING OF A VILLAGE.[1]

I WAS called, in the early part of 1878, to examine the village of Cumberland Mills, Maine, where there had been an undue amount of disease, indicating a possible defect of drainage. The village is chiefly owned by Messrs. S. D. Warren and Company, of Boston, and its population is mainly employed in their large paper mill. They had taken every measure that had occurred to them to provide in the best manner for the comfort and welfare of their people, and had expended in drains, sewers, and other sanitary appliances a very large sum; they had, in short, conscientiously done their very best, under the lights available to them, to make their village a model of healthfulness and convenience.

I found on every hand ample evidence of elaborate and costly work, of a character appropriate to the different classes of buildings. The agent's house had the usual conveniences and the usual defects of a first-class house in the city; the boarding-houses were abundantly supplied with water-works, and

[1] Reprinted by permission of Messrs. Harper and Brothers from their New Monthly Magazine.

the smaller houses had kitchen sinks with running water, cellar drains, etc.; some of the larger houses were heated with furnaces. The workmanship was generally good, and indicated that it had been guided by a good engineering skill, though quite without sanitary knowledge.

To one accustomed to the inspection of drainage works, the gravest faults of arrangement were everywhere patent. Each house had a long drain leading from its cellar to a common sewer of too large size, or to the surface of lower ground in its vicinity. Where water-closets were used, they had been erected with reference to convenience, but without reference to a proper disposal of their wastes. Most of the smaller houses had common privies adjacent to them, and in the majority of cases the drainage of the kitchen sink delivered, often through an insufficiently closed channel, into the mouth of the untrapped drain of the cellar. In some instances there were indications that these drains had become obstructed, and the discharge of the kitchen sink had overrun the cellar bottom. In other cases the foul air of the drain, or of the sewer into which it discharged, flowed back into the cellar and permeated the house. In the few instances where furnaces were used, they took their supply of cold air not from outside the house, but from the front hall, the same air being cooked over and over again — certainly with the effect of economizing fuel. The soil pipes of the water-closets were unventilated, and the insalubrity seemed to be pretty nearly in proportion to the effort which had been made to overcome it.

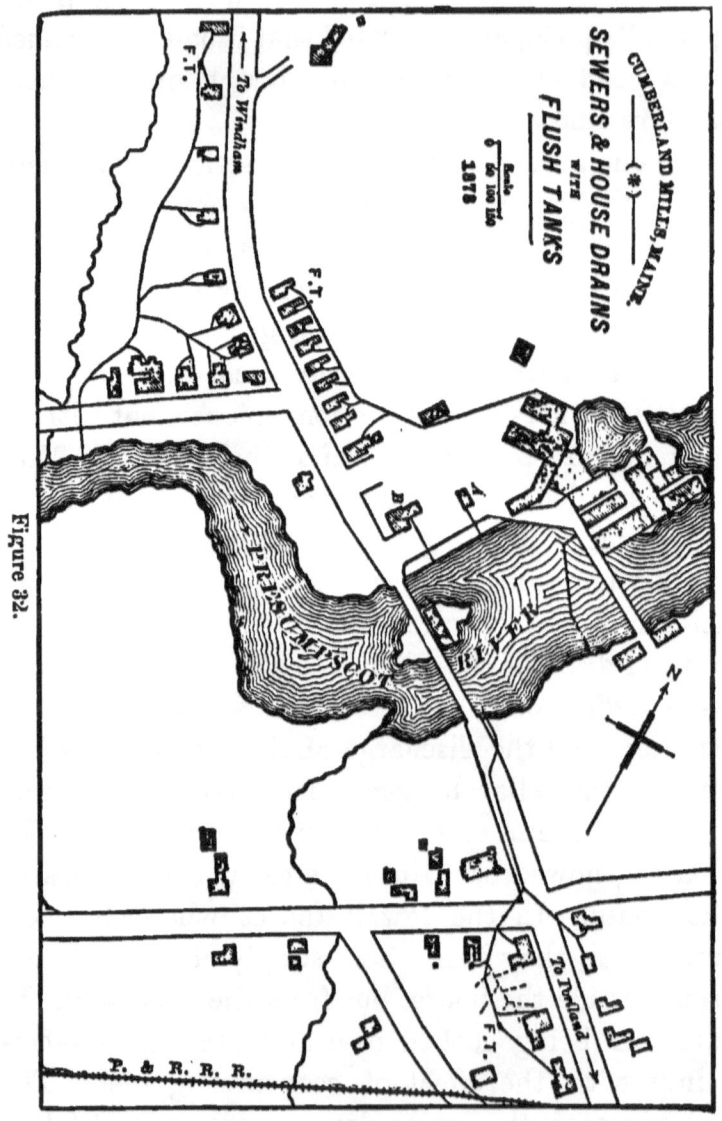

Figure 32.

I was entirely unhampered in my instructions, and was encouraged to do all that the most perfect sanitary condition required. The village lies on rolling ground considerably higher than the pond made by the damming of the Presumpscot River. This pond has a rapid and constant movement. The arrangement of the new system is shown in Figure 32. For drainage, the houses are grouped mainly into three sets, each with its independent sewer discharging into the river. *A* is the office building, where the work was very simple, and has not been changed. *B* is the agent's house, of which the drainage was entirely re-arranged, with a ventilation of its main drain and soil pipe. It is to the drainage of the operatives' houses that I desire to call especial attention.

The heavier lines indicate the main sewers, of six-inch vitrified pipe, running from the flush tanks (F T) to the river. These are laid with securely cemented joints, and with Y branches to receive the house drains, which are shown by the lighter lines. These house drains are of four-inch vitrified pipe, with cemented joints. Each one of them reaches nearly to the foundation wall of the house, and is connected under the cellar floor with the water-closet, which is in nearly every case located in the cellar. The outlet of each of the main sewers is arranged as shown in Figure 33, its extension through the bank wall of the pond and for some distance into the water being of iron pipe supported and protected by loose stone-work. At the top of the bank there is erected from a T branch of the sewer a four-inch iron pipe extending above the surface of the ground, and open at its mouth for the admission of air. There is no trap between this point and the foundation walls of the houses, each house drain being connected outside the walls with a three-inch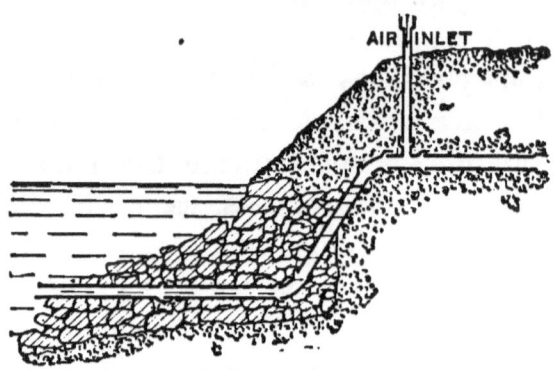

Figure 33. — Outlet of sewer, with ventilation inlet.

ventilation pipe reaching above the roof, shown in Figure 35. This arrangement secures a free circulation of air through the entire length of sewer and house drains.

At the upper end of each main sewer there is placed a Field's Flush Tank, constructed as shown in Figure 34. This is a brick chamber built in the ground, receiving in one case the drainage of a four-tenement house, and in the two others the drainage

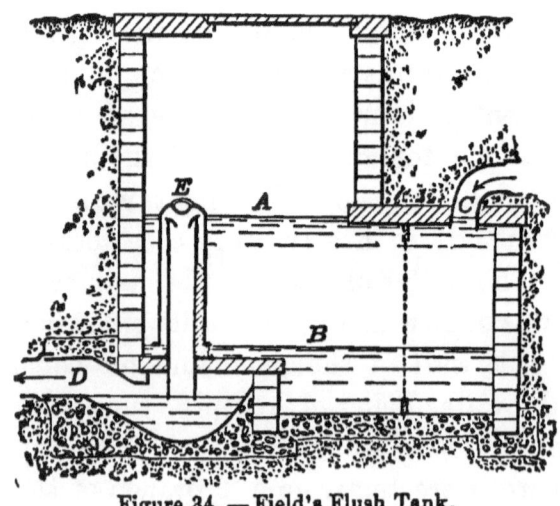

Figure 34. — Field's Flush Tank.

of the upper two houses of the series — roof water and all. The drainage enters the tank through the pipe C. A is the surface of the water when the tank is full, and B when it is emptied. The capacity of the tank between the lines A and B is about five barrels. In front of the entrance there is a wire screen to prevent the passage of coarse material. This is held in place by wooden wedges, and may easily be removed for cleansing. The depression below the line B is for the accumulation of solid matters which may not become decomposed. A portion of the tank is carried up to the surface of the ground, with a movable cover for a man-hole. E is Field's Automatic Annular Siphon, by which the tank is emptied as soon as its contents rise high enough to flow over the top of its inner (and longer) limb. The short limb is a dome inclosing the inner limb, with a water-way all around its bottom, reaching to the line B. The drainage of the remaining houses of each system flows directly to the main drain, where it may deposit more or less of its coarser matters. The drainage of the upper houses flows into the flush tank, where it is held until the top of the siphon is reached. The whole amount (five barrels) is then discharged with great rapidity into the main sewer (D), washing it clean from end to end. During storms the roof water increases this action, but the flow of sewage alone is sufficient to remove all accumulations from the sewer.

The arrangement within the houses is shown in Figure 35, where A is a tumbler tank, delivering about two quarts of water at each discharge; B is the kitchen sink; C is a check-valve trap, preventing the return of air from the water-closet to the sink; and D, the water-closet, in the cellar. The closets are of enameled cast iron, with iron traps, and iron connections with the house drains, the whole being securely set in cement, which forms the entire floor of the closet apartment. The whole cellar bottom is coated in like manner with cement. The

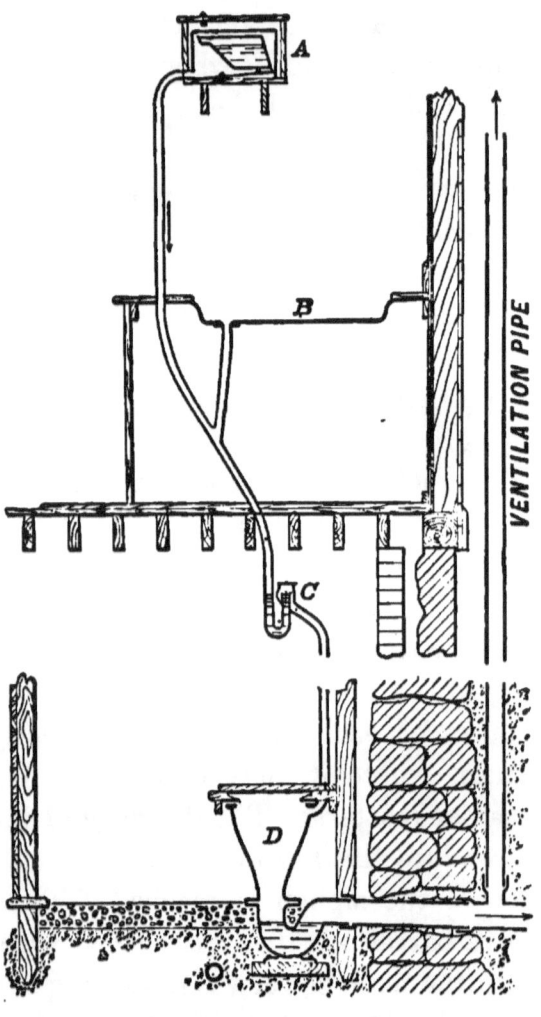

Figure 35. — House, kitchen, and cellar, with sink, water-closet, flushing arrangement, and check-valve.

closet has a wooden seat, but no riser. The whole

space around the pot is open to the air and light, and to the broom and floor cloth.

Figure 36 shows the construction of the tumbler tank, which is a small galvanized iron tank inclosed in a wooden box, of which the cover may be locked, and within which is a small faucet connected with the public water supply, and under the control of the public inspector only. Within the box, and supported on knife-edge trunnions, is a galvanized iron tumbler or tilting basin, with a capacity of about two quarts. Its normal position is shown by the solid lines (*A*), its rear end resting on a buffer of India rubber. The faucet is set to fill it at fixed intervals, usually from five to ten minutes. When nearly full, the weight of the water in the projecting lip causes it to tilt forward and assume the position indicated by the dotted lines (*B*), its front side striking an India rubber buffer, and its contents pouring rapidly out, to flow off through the outlet pipe, as shown by the arrow. When empty, its rear end is the heaviest, and it drops back into position, ready to receive another charge of water. *C* is the lock and staple by which the cover is secured. Figure 37 shows a cross section of the patent check-valve,

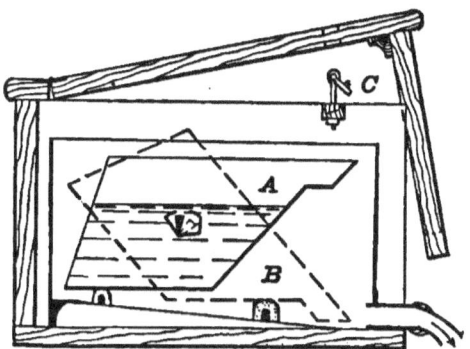

Figure 36.—Details of the tumbler tank.

by which the air of the cellar or closet is prevented from returning to the kitchen.

The frequency of the flushing discharge is a perfect security against frost; the kitchen wastepipe is kept clean, and the trapping water of the closet is renewed every five or ten minutes, day and night, all fæces and kitchen waste being carried into the drain and quite on the river before its decomposition can even begin. This frequent renewal of the water in the closet trap would be a considerable protection against foul air in the drain even were this not ventilated. In effect there is perfect ventilation only a few feet distant from the closet. The whole arrangement is entirely pure and satisfactory, and it secures the removal of all offensive waste matters in a most complete and unobjectionable manner. The same arrangements in principle are applied to the two large boarding-houses, one for men and one for women, and with equally good results.

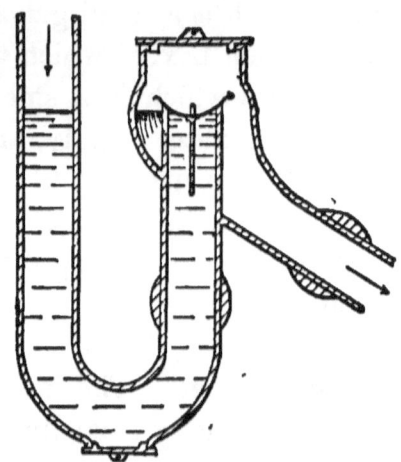

Figure 37. — Cross section of check-valve.

Other minor improvements have been made, such as the under-draining of a low tract, as shown by dotted lines near the southeast corner of the map; the removal of stables, of all pig-pens, and of all privies. Where cellars are subject to soil moisture they have been drained below the concrete, and with ample protection against the return of drain air through the old drains leading to the old sewer, or to the hill-sides. These drains have absolutely no connection with the foul-water system, which delivers below the surface of the water in the river, which is frequently and thoroughly flushed, and which is abundantly ventilated close up to the wall of every house.

Not as a matter of drainage, but as being very necessary to health, the cold air supply to furnaces, where these exist, has been cut off from the front hall registers and brought into communication with the outer air.

The houses shown on the map which are not connected with the sewers are mainly either not the property of Messrs. S. D. Warren and Company, or are to be torn down or removed.

The method of sewerage above indicated, and, so far as working-people are concerned, the method of house drainage, are almost universally applicable to country villages generally, and even to very large villages. Indeed, with a very moderate increase of size in the main sewers, where a hundred or more houses are to be drained, it is the best system available for many villages which have city charters. It would often be necessary, but by no means always, to secure some better means of sewage disposal than its discharge into a river or brook. One very important fact in this connection is apt to be overlooked, which is, that while the outflow of large and sluggish sewers is poisonous to fish, and in every way unfitted for admission to rivers, fresh fæcal matter and fresh kitchen waste are food for fishes, which are its natural and proper scavengers. The whole household drainage of a town should be carried immediately into a river by cleanly flushed sewers.

CHAPTER XII.

RECENT MODIFICATIONS IN SANITARY DRAINAGE.

IT is only about four years since the first edition of this book was published. So far as possible, I therein stated fairly the condition of the art at that time. Viewed in the light of present knowledge on the subject, parts of that first edition are already, in many respects, quite out of date. Knowledge has increased, experience has multiplied, and invention has been most fertile. The illustrations there given of the proper arrangement of house drainage (Figure 12, page 188) represented a soil pipe and drain running in an unbroken course from the sewer in the street, under the basement floor, and up through the roof of the house. Connected with it were several water-closets, a sink, and the overflow-pipes of the tank in the attic and of the service cisterns of the closets. In all cases the different vessels were separated from the soil pipe only by water-sealed traps, and only the same protection was afforded in the case of the main tank. The system thus represented is defective in several particulars.

(*a.*) The water of the tank is liable to dangerous contamination through the overflow-pipe which leads into the soil pipe, with only the insufficient protection of a water-seal, — especially insufficient as it has no certain means of renewal, and may by evaporation give direct access to the air of the soil pipe.

(*b.*) The overflow-pipes of the service cisterns may in like manner become channels for the introduction of drain air to the apartments.

(*c.*) The unprotected traps of the sink and the water-closets are inadequate for the work they are intended to perform, and all of them are susceptible, under certain conditions, of becoming empty by evaporation or by siphoning.

(*d.*) Although the soil pipe is continued through the roof, full-bore, and is open at the top, it has no provision for the admission of fresh air at its foot, which is now regarded as a matter of imperative necessity.

(*e.*) The carrying of a foul-water drain under the basement floor is to be avoided wherever possible.

These defects are sufficient, in the opinion of those instructed in such matters, to condemn this whole arrangement, which only four years ago was regarded as the best yet devised.[1]

All this indicates that the art under consideration is undergoing rapid development, and that it is by no means to be assumed that we have yet arrived at ultimate perfection in the matter.

Were I called upon to-day to specify the essen-

[1] This illustration was taken from the latest accepted English authority on such subjects.

tial features of perfect house drainage, I should include the following items: —

The establishment of a complete circulation in the main line of soil pipe and drain, allowing a free movement of atmospheric air through the whole system from end to end, together with as free a circulation through minor pipes as could conveniently be secured.

The absolute separation of the overflow of every tank or cistern delivering water for the general supply of the house from any soil pipe or drain containing a foul atmosphere.

The supplementing of every water-trap with a suitable mechanical valve, to prevent the water of the trap from coming in contact with the air of the drain.

The reduction of the size of all waste-pipes, and especially of all traps, to the smallest diameter adequate to their work.

The abolition of all brick or earthenware drains within the walls of the house, using in their stead the best quality of iron pipe, with securely caulked lead joints.

The exposure of pipes " in sight " wherever this can be done. The substitution, so far as practicable, of wrought-iron pipes for lead pipes, in the case of all minor wastes.

The coating of all iron pipes, both cast and wrought, inside and out, with " American " enamel, a glossy black coating which withstands in the most complete manner the chemical action and changes of temperature to which it is subjected in such use.

The iron pipes should be extended so far beyond the foundation of the house as to obviate the opening of joints by settlement, so common where earthenware drains are subjected to a slight movement of the foundation, or of the new fitting about it.

The object to be sought is the provision of a permanent drainage channel for the removal of all wastes, offering little asperity for the adhesion of foul matter, swept from end to end by fresh air, absolutely separated by mechanical obstructions from the interior atmosphere of the house, and literally a section of out-of-doors brought for convenience within the walls of the house, open to receive the contents of the various waste-pipes leading to it, but securely closed against the return of its air. I believe that the next step in advance will be the establishment of means by which the whole length of this drainage channel may be thoroughly flushed with clean water at least once in twenty-four hours.

As a prominent detail of house-drainage work, the long-accepted water-closet is being made the object of important modifications. The stereotyped article, the "pan" closet, has little to recommend it beyond the fact of its general adoption. It is faulty in principle, in arrangement, and in construction. While it is cleanly to look at, and lends itself readily to ornamental joinery, it has defects which should drive it out of existence. Deep down in its dark and hidden recesses, where only the ken of the plumber ever reaches, a large and sluggish trap — they call it a " cess-pool " in Scotland — is generally holding the filthiest filth in a state of offensive putrefaction. The iron chamber above this is lined with the foulest smear and slime, constantly producing fœtid and dangerous gases. The earthenware bowl which surmounts this is set in putty, which yields to corrosion and to the jar of frequent use, until it leaks foul air, often in perceptible quantity. The panful of sealing water soon becomes saturated with foul gases, which exhale thence into the house. The whole apparatus is incoffined in tight-fitting carpentry, which shuts in the leakings and the spatterings and their vapors from the free access of air, boxing up in the interior of the house, and generally in free communication with the spaces between the walls and under the floors, an atmosphere heavy with the products of organic decomposition, and faintly suggestive to (the unwonted nostril) of the *mus decumanus defunctus*.

Some of these defects were recognized and pointed out in my earlier papers. I then believed that the difficulties of the case had been solved in great measure by the Jennings closet. It now seems that this closet and the whole class to which it belongs are seriously defective; and, in the absence of anything better, I am disposed to go back to the simple " hopper " closet (Figure 38), such as is used in the cheapest work, and to depend on frequent and copious flushing to keep it clean. This closet has the great advantage that its only trap is in sight at the bottom of its pot. There is no inner "chamber of horrors" concealed by a cleanly exterior. I have recently used a number of these closets supplied with various sorts of apparatus for periodical flushing, and I find that wherever a half-gallon flush can be given every ten or fifteen minutes they are kept perfectly clean. I have no doubt that flushing every half hour would keep them free from all sanitary objection. This would require a supply of about twenty-five gallons *per diem*.

Figure 38. The hopper closet.

Recent invention has been turned in the direction of the provision of mechanical appliances for separating the trapping water from the air of the soil pipe or drain. There are several devices which accomplish this purpose, — one of them (Figure 37) my own, and more than one of them constituting a very great improvement upon, and indeed an absolute step in advance of, anything in use five years ago.

Another most important matter of recent development is the through and through ventilation of soil pipes. Formerly the soil pipes invariably stopped at the highest closet of the house. When the danger of *pressure* came to be understood, it was considered imperative in all work of the best class to carry a vent-pipe out through the top of the house. As this pipe, from the smallness of its size and from the irregularities of its course, had but limited capacity of discharge, the necessity was quite generally recognized for carrying up the soil pipe itself, full-bore, through and above the roof. This was the point reached at the time of my earlier writing. It soon became evident that even this extension of the large pipe afforded no real ventilation. A deep mine shaft cannot be ventilated by simply uncovering its top. No complete frequent change of air can be effected in a soil pipe by merely opening its upper end. Air must be introduced at the bottom to take the place of that which is discharged at the top. It is now considered imperative in all good work to open the soil pipe at both ends, or at least to furnish the lower part of the pipe with a sufficient fresh-air inlet to effect a thorough ventilation of the whole channel.

We have heard so much of "sewer gas" that we were in danger of ascribing the production of this foul air only to the sewer and cess-pool. Indeed, the majority of sanitarians to this day seem to believe that if they can effect a thorough disconnection between the sewer or drain and the waste-pipes of the house they have gained a sufficient protection against sewer gas. The fact is that that combination of the gaseous products of organic decomposition which is known by the generic name of sewer gas is very largely produced by the contents of the house-pipes themselves. Not only in the traps, where the coarser matters accumulate, but all along the walls of the smeared pipes, where filth has attached itself in its passage, there is a constant decomposition going on which is producing its constant results. The character of this decomposition and the character of the produced gases are greatly influenced by the degree to which access is given to atmospheric air. The more complete the ventilation, the greater the dilution of the gases formed and the more complete their removal, and also the more innocuous their character. Under the most favorable circumstances, the contained air of a soil pipe must be offensive, and is likely to become dangerous; so that, however thorough the ventilation, we must still adopt every safeguard against its admission into the house. The facility with which foul gases penetrate water and escape from it makes the water-seal trap, which is now our almost universal reliance, an extremely inefficient protection. There can be no real safety short of the adoption of some appliance which shall keep every outlet securely closed against the possible return of drain air.

Mr. Elliot C. Clarke, the principal assistant engineer in charge of the Improved Sewerage Work of Boston, in a paper entitled "Common Defects in House Drains," contributed to the Tenth Annual Report of the Massachusetts State Board of Health, says on the subject of sewer gas: " The writer has no wish to be an alarmist. The risk from sewer gas is probably not so great as many suppose; it is a slight risk, but a slight risk of a terrible danger. If a man thinks there is no need of insuring his house because his father lived in it for fifty years without a conflagration, he has a right to his opinion." Professor Fleeming Jenkin, in his "Healthy Houses," says, "Simple sewer gas is little worse than a bad smell. Tainted sewer gas may be so poisonous that a very little introduced into a bedroom — so little as to be quite imperceptible to the nose — shall certainly give typhoid fever to a person sleeping there. The germ is a spark, the effects of which may be unlimited. We do not content ourselves with excluding the great majority of sparks from a powder magazine; we do our best that not one may enter."

While the water seal is very defective in itself, it is a very important adjunct to any mechanical means of separation that may be adopted, and all necessary precautions should be taken to prevent its removal by "siphoning," — the sucking out of the water by the partial vacuum caused by the flow of water in the main pipe, to which its outlet leads. To prevent this siphoning action often taxes the ingenuity of the engineer more than any part of house-draining work; and until special devices are made to meet the exigency this must remain the most difficult and intricate part of the house drainer's task.

Any one whose attention is given to sanitary work must be more and more struck with that peculiarity of human nature which assures us of the exceptional excellence of our own belongings. I have rarely been called to examine the drainage of a house without being told that I was sent for merely as a matter of *extra* precaution. I have never completed any examination without discovering serious sanitary defects, — not merely such errors of arrangement as were universal until a short time ago, but actual, palpable bad condition, which the owner and his plumber at once acknowledged as of a grave character. Leaks in drains under the cellar floor, or in or near the foundation; lead waste-pipes eaten through by rats, and spilling their flow under the house; lead soil pipes perforated by corrosion; imperfect joints leaking drain air within the partitions; the accumulation of dirty sloppings under the bench of the water-closet; and even untrapped connection between some room and the soil pipe, or the direct pollution of the air over the tank through its overflow-pipe, — these are most common faults, and some one of them I have found to exist wherever I have looked for them in a " first-class " house, where it was naturally supposed that the most perfect conditions prevailed.

In no department of sanitary work has progress been more marked than in the improvement foreshadowed on page 196 *et infra*, concerning the disposal of the liquid wastes of country houses by the process of sub-surface irrigation. Like all radical improvements, it has had its share of prejudice to overcome, and it by no means found the professional public ready to accept it as the demonstrated success which English experience had shown it to be. It is now quite safe to say that, among all engineers and architects who have given attention to the matter, it is acknowledged to afford the best solution yet attained of this most difficult problem. I know very many cases of its adoption, often without professional guidance, and carried out in a rule-of-thumb sort of way, and I have heard of none that is not satisfactory. It does away with that king of nuisances, the cess-pool, and disposes of all manner of liquid waste insensibly, completely, and safely. The credit for this improvement is due primarily to the Rev. Henry Moule, the inventor of the earth-closet, and hardly less to Mr. Rogers Field, C. E., who relieved it of its chief embarrassment by adapting to it his automatic flush tank. This system has recently received the unqualified indorsement of that highest American sanitary authority, the Massachusetts State Board of Health, which in a circular issued in April, 1879, says: "Chamber slops, and slop water generally, should never be thrown on the ground near houses. They may be .... used by distribution under the surface of the soil in the manner described on page 334 of the Seventh Annual Report of the State Board of Health, and now introduced in the town of Lenox, Massachusetts. . . . If water-closets are used, and there are no sewers, the best disposal of the sewage is by the flush tank and irrigation under the surface of the soil, as described on page 135 of the Eighth Annual Report of the State Board of Health."

This system has been for two years in full operation for the entire sewage of the village of Lenox, where it has proved itself an absolute and unquestionable success. The question which seems to arise in every Northern mind when this method is suggested relates to the possible effect of severe frosts. It seems now to be clearly demonstrated that this consideration may be left entirely out of the account, no instance having been cited of the least obstruction from this source. This point will be more fully treated farther on.

The progress made in the matter of town drainage has not been less than that in the twin department of house drainage; but the advance has been thus far — at least so far as this country is concerned — more a matter of theory than of practical application, and it relates more to villages and to what may be called village-cities than to larger places, like New York, Boston, and Philadelphia.

Sewerage was long confined to large towns, and it reached its development under the direction of engineers trained to foresee all possible contingencies, and to pitch their work on a scale adequate to cope with them. Having usually ample means at their command, and with the inclination to work after great models, their works have generally been costly and vastly comprehensive. So far as the drainage of the great cities is concerned, there is much to be said, too, on the other side, and it has been ably said. My present purpose relates chiefly to the sewerage of villages and country towns having a considerable proportion of uncovered and unpaved area. There are hundreds of towns in this country sadly in need of draining, which cannot afford the gigantic and costly work of introducing such a system of sewers as it is uusal to find in a great city. Quite generally, when the question of their drainage arises, a city sewerage engineer is consulted, and a plan is prepared which remains unexecuted because of its excessive cost. By far the larger part of this cost is due to the fact that the proposed system contemplates the drainage of such sub-cellars as are rarely found in country towns, involving a depth that would probably never be needed, and the removal of the storm water, which, after the area shall have become covered and paved, might flow off by the public sewers. It would be better, in the case of all rural towns, to disregard the question of storm water entirely. This may be more safely and much more cheaply removed over the surface. The only reason for admitting it to the sewers would be to prevent injury to property. Under the circumstances we are considering, the danger of this is not sufficient to justify the expense; nor is it sufficient, were there no question of expense, to justify the sanitary and economical disadvantages of providing for it by a system of large sewers. It is better to keep above ground, and to discharge by the natural means of outflow, all water which may be so disposed of without offense or danger to health, — that is, all or nearly all rainfall. The extent to which the first flow over a paved road-way may properly be admitted to the sewers is a question to be decided according to the circumstances of each case. It is generally wiser to keep such road-ways clean by sweeping than to use the rain-fall as a scavenger.

What towns of the class under consideration need — and they need it very imperatively — is a perfect means for the removal of the foul wastes of households, factories, etc., and for the draining of the sub-soil, if this is springy or wet. They should only be called upon to spend the money necessary to secure these ends. If they can learn to limit their demands to this absolute requirement, their sanitary improvement need no longer be the bugbear that it now is.

The advantages of small-pipe sewers have been sufficiently stated, except, perhaps, with reference to the single matter of ventilation. It is much easier and more simple to secure the needed change of the atmosphere of a small chamber than of a large one, and the usual means, which are but partially effective in the case of a large brick sewer, are ample for the complete ventilation of a small pipe. Hitherto the objection has held, in the case of pipe sewers of less than ten inches in diameter, that when they become obstructed it is a difficult and costly matter to clear them. But for this objection, there was no reason why six-inch sewers might not be used for all villages or parts of towns having a population of not more than one thousand; for a six-inch pipe laid even with a very slight inclination has ample capacity for the discharge of all the household waste of such a population.

We have now reached the point where there is no reason whatever to apprehend the obstruction of such a sewer by anything that can get into it through proper and properly arranged branch drains. Rogers Field's Flush Tank, as arranged for the periodic flushing of such sewers, may be confidently relied on to keep them swept clean of everything that may enter them. The accompanying diagram (Figure 39) shows the construction of the annular

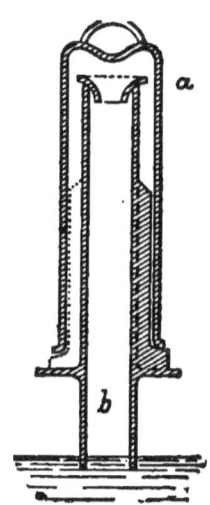

Figure 39. — Rogers Field's Annular Siphon.

siphon, which is the essential feature of this tank. A siphon of this form, four inches in diameter, comes into action with certainty under a stream of one tenth of a gallon per minute; so that a tank having a capacity of one hundred and fifty gallons, placed at the head of each branch sewer and fed by a stream which will fill it once in twenty-four hours, will give it a thorough daily flushing, and keep it clear of all obstructions. No matter how limited the public water supply may be, this small amount can always be spared for the work. Where there is no public supply and no available extrinsic source of flushing water, the sewage itself from a few of the upper houses along each lateral sewer, together

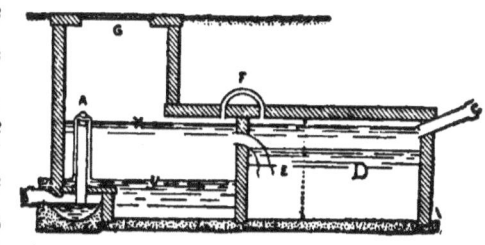

Figure 40. — Rogers Field's Flush Tank for Sewers.

with their roof water, may be collected in the tank and used for the flush.

This simple device has proved itself, both here and in England, to be entirely reliable and effective. It may safely be assumed that it has secured a reduction of the cost of the drainage of small towns to one half of what was formerly necessary.

It has been held hitherto to be one of the advantages of sewerage that the imperfect joints or imperfect material of the sewers afford an outlet for superabundant soil water, and secure a valuable sub-soil drainage. It is coming to be understood that the same channels which admit soil water to the drain will admit drain water to the soil, robbing the sewers of the vehicle needed for the transportation of their more solid contents, and causing a dangerous pollution of the ground, of cellars, and of drinking-water wells. The foul-water sewers should be as absolutely tight as the best material and the best workmanship can make them, and the drainage of the ground should be effected by the use of agricultural drain-tiles, constituting an entirely separate system, which, while they may for economy's sake generally occupy the same trenches with the sewers, should be carefully arranged to prevent sewage matters from entering them.

The question of sewage disposal is the great unanswered question of the day. We are familiar with the objections to the methods usual here. European countries, which have been forced by the density of their population to give especial attention to this subject, have as yet hardly got beyond the point of proving that there is no royal road to success, and that whatever theory may say on the subject, sewage not only has no value to the community producing it, but it cannot be got rid of except at considerable cost.

The only method thus far developed which is entitled to consideration here, aside from discharge into the sea or into a running stream, is purification by application to the soil, with or without the agricultural consideration. Whether by surface irrigation, by the use of sub-surface absorption drains, or by intermittent downward filtration, this method of its disposition, properly applied, is absolutely complete and satisfactory. The opinion has quite naturally prevailed that the severity of our winter climate debarred us from availing ourselves of it. The experience of the severe winter of 1878 9 has fully justified the opinion of those who have maintained that this objection is not a real one. In England the sewage-irrigation farms have taken charge of the effluent without interruption throughout a season of almost unexampled severity. At Berlin a like immunity has continued throughout the winter; and even at Dantzic, near the mouth of the Vistula, in a climate nearly as severe as that of St. Petersburg, and where provision was made for a direct discharge into the river during the winter season, the disposal by irrigation is said, to the surprise of all, to continue uninterrupted in the coldest weather.

At the Nursery and Child's Hospital on Staten Island, winter. overtook us before our absorption drains could be laid. The flush tank, which holds one day's sewage, was made to discharge over a low spot near the absorption ground. Even in the coldest weather the entire outflow settled away into the earth before the next flood was delivered. Evidently the warmth of the sewage is in all cases sufficient for it to thaw its way into the ground. This is, without doubt, the explanation of the continued working of the shallow drains under my own lawn during nine consecutive winters, although at least once the ground was frozen to a depth of two and a half feet below them.

INDEX.

Ash-closets, 214, 271, 275, 280.
Ashes for earth-closets, 250.

Batavia (N. Y.), removal of malarious condition, 109.
Bath-tubs, 103.
Blunt's overflow for Jennings's closet, 79.
Boards of Health of England. General conclusions as to sanitary drainage, 118.
Boards of Health should decide what may be admitted to the sewer, 117.
Boards of Health should regulate the sites of houses, 12, 57.
Boston, its drainage, 64.
Brick sewers, objections to, 146.
Brierly, typhoid at, 40.
Broad Street pump, 68.
Brooklyn, amount of sewage, 129.

Capacity of sewers, 134, 136, 138.
Causation of disease (Dr. Simon), 35.
Cellar drainage, 74.
Cellars, 73, 83, 84.
Cellars, flooding of, 128.
Cement pipes for sewers, 145.
Cerebro-spinal meningitis, 53, 96, 101.
Cess-pool, 79, 80, 89, 107.
Charcoal ventilators, 159.
Chimneys and flues for sewer ventilation, 156.
Chimneys, house-drain ventilators in, 93.
Choking of sewers, 133.
Cholera, bad drainage, 53.
Cholera, mortality from, 20.
Cleansing sewers, 165.
Closets, water, 78, 79, 190, 193.
Construction of sewers, 180.
Consumption from moisture, 44.
Contagions and miasms, 57.
Contamination of soil, 82.
Contamination of wells, 81.
Country house-drainage, 79.
Crosshill, typhoid at, 40.
Croydon, cost and value of sewerage, 115.
Croydon, typhoid and sewer ventilation, 109, 112, 151, 157.

Dampness, 48, 73.
Dams for flushing sewers, 167.
Death-rate in London at different periods, 21.
Death-rates in England, 18.
Decomposing matter in cellars, 83, 84.
Denton and Field's sewage tank, 319.
Deodorizing, 91.
Deposits in sewers, 144.
Derby, Dr., without filth typhoid is not born, 56.
Details of house drainage, 186.
Diarrhœa, 96.
Diphtheria, 53, 54, 55, 96, 101.
Diphtheria, and sewer gas, 44.
Discharge of sewage at different houses, 133.
Diseases from sewer gas, 96.
Disease, stamping out sources of, 88.
Disinfectants, not reliable, 57.
Disinfectants, objectionable, 165.
Disinfection, 90, 91.
Disposal of house slops, 196, 329
Disposal of sewage, 314.
Drainage, as affecting fever and ague, 45, 48, 50.
Drainage entails responsibility, 79.
Drainage, from kitchen sinks, 75.
Drainage of houses, 186.
Drainage of towns, 105.
Drains and sewers, sizes of, 135.
Drains, dangers of, 91.
Drains, private, 106, 177, 206, 208.
Drains, ventilation of, 91, 95.
Drinking-water, 80, 81, 82, 83.
Drinking-water, infected, 26, 27, 66, 67, 68.
Dry conservancy system, 213.
Dry earth, 90-93.
Dry-earth system, 68, 213.
Dysentery, 53.

Earth-closet apparatus, 244.
Earth-closet, ashes may be used in, 250.
Earth-closet, hygienic advantages, 263.
Earth-closet, manure from, 225, 234, 258.
Earth-closet, management of, 217.

364 INDEX.

Earth-closet, Netten Radcliffe's report, 253.
Earth-closet, necessary control, 225, 260.
Earth-closet system, 214.
Earth for closets, how obtained, 219.
Earth, kinds to be used in closets, 217, 219.
Earth privies, how arranged, 250.
Earth system for towns, how managed, 224.
Earth, the same may be repeatedly used in closets, 218.
Eastwick, sewerage of, 325.
Effete matters dangerous, 79.
Egg-shaped sewers, 147.
English theory of typhoid infection, 24.
Epidemics from bad sewerage, 118.
Excrement as a vehicle of typhoid contagion, 152

Factories and other industrial establishments should pay for their own extra sewer requirements, 116.
Fall of sewers, 142.
Fergus, Dr., 96.
Fever and ague, Staten Island, 45.
Field's flush-tank, 198.
Filtering through soil, 81.
Filth, hiding it under ground, 88.
Flint, Dr., on typhoid, 37.
Flow in sewers, how retarded, 184
Flushing, 106, 127, 165.
Flush-tank, Field's, 198.
Forms of sewers, 146.
Foul drainage of houses, 75.
Foundations, 73.
Friction in sewers, 142.

Goux system, 214, 265
Grease traps, 77, 195.
Gutters cheaper than sewers, 128

Health, effect of sanitary work, 20.
Health question, its financial aspect, 51.
Holland, Liernur's system in, 305.
Hotel, Grand Union at Saratoga, 141
House drains, dangers of, 91.
House drains, how to be laid, 76.
House drains, obstructions in, 76.
House drains should enter sewers above water line, 164.
House drains, sizes for, 187.
House drains, ventilation of, 91-95.
Householder, problems to solve, 72.
House slops, disposal of, 196, 329.
Houses, drainage of, 71, 186.
House sewerage, 75.
House-slops, disposal of, 196
House ventilation, 201.
Hull privy, 279.
Hull system, 276

Hygienic advantages of earth system 263.

Impervious courses in foundations, 73
Industrial establishments should pay for their own extra sewer requirements, 116.
Inspectors of sewage work, 183.
Intercepting sewers, 124.
Intermittent water supply as a cause of typhoid infection, 41.
Irrigation, 315.
Irrigation tank, sewage, 319.

Jennings's closet, 78, 79, 193.
Johnson, Prof. S. W., on soil pollution, 24.
Junctions of sewers, 177

Kitchen drain, 75.

Lamp-holes, 175.
Large sewers wasteful, 118.
Latham's charcoal ventilator, 161.
Lead pipes, decay of, 95, 96.
Leconte, Dr, recommendations, 34.
Leconte, Dr., report on typhoid at St. Mary's Hall, 33.
Lewes, typhoid from intermittent water supply, 41.
Liebermeister on typhoid, 58.
Liernur's system, 284.
London pumps, 27, 58.

Malaria, 108, 109.
Malaria makes all diseases more serious, 116.
Malarious condition made worse by population, 110.
Man-holes, 175.
Manure question as affecting earth system, 225.
Manure from earth-closet, 234.
Maplewood School, 30.
Miasmas and contagions, 57.
Mists deleterious to health, 47, 48.
Mortality by war and by disease, 17.
Moule's apparatus, 244.
Moule's dry-earth system, 214.
Moule's system for house slop disposal, 196.

Nervous toothache, bad drainage, 53.
Netten Radcliffe on the earth system, 253.
Neuralgia, bad drainage, 53.
Newport death-rate, 61.
Newport, sanitary drainage, 59.
New York, cost of cleansing sewers by flushing, 170.
New York, cost of cleansing sewers by hand, 170.
New York, old water-courses not connected with diphtheria, 58.

INDEX.

New York rain-fall, 132.
Nuisances, 87.

Odor and taste of dangerous drinking water, 83.
Odor of sewer gas, 94.
Odor not a sufficient test of poisonous condition, 108.
Organic matter, refuse, 87
Outlets, 124.
Outlets closed by tide, 125.
Outlets exposed to winds, 126, 156.
Over-Darwen, typhoid at, 42.

Pail system, 273.
Palmer, Dr., report on "Maplewood" fever, 31.
Pan closet, 192.
Pettenkofer's theory of typhoid infection, 23.
Pipe sewers need flushing, 173.
Pittsfield (Mass.), "Maplewood fever," 30.
Plans of sewerage, 123.
Plumbing arrangements, 187.
Plumbing defects, 73.
Plumbing requires careful supervision, 15.
Pneumatic emptying of vaults, 275.
Pneumatic system of drainage, 70.
Pneumatic system, Liernur's, 284.
Pressure of air in sewers, 154, 156.
Private drains, 106, 177, 206.
Private drains, how to be laid, 76.
Private drains of brick or stone, 118.
Private sewers, 106.
Privies, 272.
Privies, earth, how arranged, 250.
Privy vault, 90.
Providence rain-fall, 131.
Providence, rules for laying private drains, 208.
Providence, sewerage well managed, 181.
Public sanitary improvement, 14.
Purification of sewage, 318.

Rain-fall, 128, 130.
Rain-water necessary in sewers, 150.
Rain-water pipes, as sewer ventilators, 156.
Removal of sewage should be immediate, 113.
Rochdale system, 273.
Rome, malarious region about, 109.
Rowe on storm-water discharge, 132.
Rules for laying private drains, 208.

Sand foundations, 73.
Sanitary dangers, 87.
Sanitary improvement in British navy, 22.
Sanitary works, effect on health, 20.
Saratoga sewer, 139, 184.

Scarlet fever, 96.
Scarlet fever, bad drainage, 53.
Scavenging, 88–90.
Separate system, 149.
Sewage discharged into tidal waters, 127.
Sewage, disposal of, 314.
Sewage, effect of vegetation on, 200.
Sewage irrigation, 149, 315.
Sewage not dangerous when fresh, 112.
Sewage, pumping, 118, 124, 126.
Sewage purification, 318.
Sewage, quantity of, for each person, 129.
Sewage, recent, floats, when macerated, sinks, 127.
Sewage tank, 319.
Sewage with and without water-closet discharge, 113.
Sewerage, 123.
Sewerage and water supply should be provided at the same time, 108, 114.
Sewerage, cost and value of, in Croydon, 115.
Sewerage, how to be paid for, 115.
Sewerage should be comprehensive 115.
Sewer gas, 94.
Sewer gas and diptheria, 44.
Sewer gas, dilution of, 158.
Sewer gas, how admitted to the house, 98, 99, 103, 189.
Sewer gas transmitted through the water of traps, 97.
Sewer junctions, 177.
Sewer pipes, 144.
Sewer ventilation, 104, 149.
Sewer work must be thorough, 181
Sewers, capacity of, 134, 136.
Sewers, choking of, 133.
Sewers, construction of, 180
Sewers, deposits in, 144
Sewers, fall of, 142.
Sewers, flushing of, 165.
Sewers may become elongated cesspools, 101.
Sewers, pressure of air in, 154, 156.
Sewers, private, 106.
Sewers, requirements of good, 105.
Sewers, safe, 152.
Sewers, sanitary authorities should decide what is to be admitted to them 117.
Sewers, sizes of, 127, 135, 138, 143.
Sewers sure to convey contagion, 101, 112.
Shawneetown (Ill.), formerly malarious, 110.
Shedd, J. Herbert, plan of street gullies, 175.
Sickness, proportion to deaths, 17.
Simon, Dr., on "ferments of disease," 35.

INDEX.

Simon, Dr., on the action of infective matters, 37.
Simon, Dr., on water-closets, 190.
Sink drain, 75.
Site of house, its relation to typhoid, 56.
Smith's closet, 78, 79.
Smith, Dr. Stephen, "Visitation of Providence," 68.
Soil moisture as a cause of consumption, 44.
Soil moisture and fever and ague, 45.
Soil moisture cause of dangerous mists, 47.
Soil pipes, 94.
Soil pipes as sewer ventilators, 163.
Soil pipes, inspection of, 97.
Soil pipes unventilated, 100.
Soil pipe ventilation, 194.
Soil water drainage, 108.
Stagnant pools, 110.
Staten Island, fever and ague, 45.
Storms, records of, 129, 131.
Storm-water removal, 124, 128, 132.
Street gullies, 173.
Subsoil irrigation with house slops, 196.
Sulphuretted hydrogen, 153.
Surface water, 80.

Taste and odor of dangerous drinking-water, 88
Taunton, "the cleanest town in England," 63.
Thoroughness of sewer work, 181.
Tide valves, 178.
Tide water outlets, 125.
Town sewerage, 123.
Traps, forcing of, 98, 99, 102, 126, 189.
Traps for grease, 195.
Traps, siphoned out, 99, 103.
Traps transmit gases, 97, 189.
Trough closets, 282, 283.
Tumbler closets, 282.
Tumbler for flushing sewer, 167.
Typhoid, 101.
Typhoid and sewer ventilation, 153.
Typhoid at Brierly, 40.
Typhoid at Crosshill, 40.
Typhoid at Lewes, 41.
Typhoid at Over-Darwen, 42.
Typhoid at Uppingham, 39.
Typhoid, contagion by means of excrement, 152.
Typhoid, English theory, 24.
Typhoid first attacks those who have become debilitated by foul surroundings, 57.
Typhoid from infected milk, 40, 41.
Typhoid from infected water, 26.
Typhoid from unventilated sewers, 151.
Typhoid infection from intermittent water supply, 41.

Typhoid, its causes controllable, 29.
Typhoid, its contagion, 36.
Typhoid, its relation to site of house, 56.
Typhoid, Liebermeister on, 58.
Typhoid may be produced *de novo*, 30.
Typhoid, Pettenkofer's theory, 23.
Typhoid propagation, 25.
Typhoid rate, reduction by sanitary works, 21.

Uppingham, typhoid at, 39

Vaults, 272.
Vaults, privy, 90.
Vegetable decay as a cause of typhoid, 29.
Vegetation, effect of, on sewage, 200.
Velocity of flow, increase of, in sewers, 134.
Velocity of flow in sewers, 134, 143.
Ventilation of house drain, 91, 93, 95
Ventilation of houses, 201.
Ventilation of houses, effect of, on health, 206.
Ventilation of sewer, 104, 105, 106.
Ventilation of sewers by frequent opening into streets, 158, 162.
Ventilation of sewers through soil pipes of houses, 163.
Ventilation of soil pipes, 100, 194.
Ventilation of water-closets, 78.
Village sewers, 138, 139.
Voelcker, Dr., on earth-closet manure, 234.

Wash-basin, 103.
Wash-basins (stationary), 78.
Water carriage, 118.
Water-closets, 190.
Water-closet, defects in, 77.
Water-closet, Dr. De Chaumont on, 192.
Water-closets, Dr. Hill on, 192.
Water-closet ventilation, 78.
Water-closets should always be supplied from separate cistern, 191.
Water supply and sewerage should be provided at the same time, 108, 114.
Water supply, infected, as a cause of typhoid, 43.
Water-closet system has increased the prevalence of certain diseases, 101, 111.
Water-closet unsuitable for poorer classes, 191.
Water traps, 93.
Wells, 80.
Williamstown (Mass.), typhoid from poisoned well, 67.
Windsor (England), typhoid from defective sewer, 112.
Worthing death-rate, 62.
Worthing, sanitary drainage, 62

www.ingramcontent.com/pod-product-compliance
Lightning Source LLC
Chambersburg PA
CBHW020235240426
43672CB00006B/541